Weather Map Handbook

Second Edition

A guide to the Internet, modern forecasting,
and weather technology

Tim Vasquez

Weather Map Handbook
Second edition
Published October 2008

ISBN 0 9706840 7 X

Printed in the United States of America

Weather Graphics Technologies
P.O. Box 450211, Garland TX 75045
(800) 840-6280 fax (206) 279-3282
Web site: www.weathergraphics.com
servicedesk@weathergraphics.com

Contents

Acknowledgements

This book, compared to other titles I have written, was a relatively solitary effort. Thanks to Chuck Doswell for assistance on pinning down the source of an older stability index, to Greg Stumpf for clarifying information about NSSL's WSR-88D algorithm developments, and most of all to Shannon Key for helping with several nomograms and proofreading. I also appreciate some insight contributed by Lon Curtis, Stephen Corfidi, John Monteverdi, Paul Markowski, David Blanchard, Patrick Kerrin, and Tim Marshall.

This book attempted to bring in all available meteorological charts, and I would love to hear about omissions so that this can be updated in the next edition. Please submit suggestions and corrections to <*servicedesk@weathergraphics.com*>.

Introduction

Weather aficionados, a group which includes just about anyone predisposed to picking up this book, are living in an awesome science-fiction tale. In Chapter One, which began in 1983, the only economical weather data came from NOAA Weather Radio, where a hobbyist would spend ten minutes listening to the voice broadcast to find bits and pieces about what was happening in the towns nearby. For $1000 per month there was the NOAA Weather Wire, which wasn't quite as good as the upscale FAA 604 data feed. Or if you had a costly home computer, you could pay as you go — $1 per minute plus long distance charges would buy downloadable data at 1200 baud from a private vendor. Electronic weather maps could be downloaded as very coarse images in Tektronix 4010 format, viewable only on a good PC with a proprietary terminal software package. For the rest of us without a big wallet, there was CompuServe's Weather forum, which offered the gamut of North American raw observations for about 50 cents per session plus connect charges.

Chapter Two began in earnest in 1995. The technological magic of the Internet and cheap technology have brought the global weather telecommunications infrastructure to every person's desktop. It's possible to obtain almost every shred of global weather data at lightning-fast speeds. Vast graphics storehouses from models and weather display programs are available at the click of a mouse. What do you do with these incredible resources?

It is my hope that this book provides an excellent introduction for the enthusiastic novice as well as a valuable reference book for seasoned hobbyists. At the very least it is designed to help guide readers through the vast sea of weather data which is getting deeper every day. The pitfalls, the strengths, and the quirks of each product are pointed out, and informal conventions such as coloring and symbology are defined.

It's an exciting time to enjoy the field of meteorology! Technology simply has not caught up with the complexity of atmospheric physics. Every five years, most weather agencies pour millions of dollars into major computer upgrades. The bureaucrats tout them as a panacea, but in reality all they do is give us bigger and better tools. Regardless of the quality of tools that are produced, the challenging job of forecasting must still be done. This leaves plenty of room for the person who wants to learn the art and mystery of forecasting using 21st-century technology: applying experience, intuition, and pattern recognition from past weather events. A background in differential equations and a college degree are not necessary to make an accurate forecast. All it takes is the desire to participate and the enthusiasm to tackle an advanced subject when things are unclear. Little by little one can become a master at this sophisticated, captivating area of science. It's the ultimate puzzle!

TIM VASQUEZ
October 2008
Norman, Oklahoma

Basic Forecasting Concepts

The world is at your fingertips. The high-speed cablemodem is toasty, the lights flickering confidently. Thousands of charts are at your disposal, much more data than a typical National Weather Service office could dream about in 1990. Yet you can't help but to sit there and ask: What do I look at?

What's your objective?

Certainly you could browse randomly through dozens of Internet weather charts. In fact every time you sat down you could pull up each and every chart listed in this book! However, this is inefficient and does little to provide you with meaningful information. There is no way the human brain can sift coherently through hundreds of charts.

The question you must ask yourself is *what am I trying to accomplish?* The answer, combined with a rough understanding of meteorology, will help you seek out the ingredients and patterns you are looking for, and in turn this book will help you figure out what charts you should concentrate on.

For example, if you are sailing, are you trying to find the best wind patterns? If so, you'll be focusing on surface charts, and will also examine upper-level charts for features that could locally strengthen the surface pressure gradient. Are you a chaser expecting storms to fire any minute? The ingredients that would favor initiation in a specific location would include convergence and the presence of cumulus towers. Thus you will be looking at small-scale surface plots and visible satellite imagery.

Picking a good web site

Your next challenge will be deciding what web sites work best for you. It's always best to research your options on a fair-weather day. There's nothing worse than scrambling to find the best radar chart when a historic storm is about to punch through the state!

How timely are the updates? Your favorite site might have the best graphics, but if 06Z rolls around and the RUC isn't yet posted, it's a good idea to have a backup in mind.

Are you getting the fields and detail you want? Do you want wind barbs with that RUC surface prog? If so UCAR or NCEP might be best, otherwise College of DuPage or Unisys might be your choice. Are you happy with stick-figure geography or do you like cool, antialiased borders with terrain? Do you like the power of multiple fields, or is simplicity your cup of tea? Look around!

Do the charts print well? If you love making printouts and posting them on your bulletin board, charts with black backgrounds will look atrocious and will drain your ink-cartridge budget. See if the website offers a "printer friendly" option for each map. This provides the same graphic with a white background.

Will software products do a better job? Depending on what kind of chart analysis you are doing, Digital Atmosphere (the author's plotting software), RAOB, Sharp, and other standalone programs may be much more suited to the task at hand and give more professional results. Dabble a bit with each of them.

Are there newer sources? Weather charts on the web change monthly. Listen in on forums such as storm2k.org, stormtrack.org, and WX-TALK. Examine other people's bookmarks to see what they are using. Check in on sites you don't typically use. Even some of the Web sites in this book may be obsolete after several months. If you get a "file not found" error you can often find the new charts by deleting the right site of the URL up to the last solidus, trying again. Keep shortening the URL in this manner until you get to a working page.

Bookmark your favorite sites. There's nothing more aggravating than trying to remember that obscure website that gave you an excellent weather chart. Maintain a special bookmark folder and add to it as you find new charts you like.

Weather basics: back to school

Although a complete discussion about forecasting is far beyond the scope of this book, a brief overview of forecasting process is warranted. The overwhelming question that most forecasters try to solve daily is, "Where will there be clouds and precipitation?"

Obviously the clouds and precipitation may not have formed yet, so looking for the underlying causes, rather than what's there already, is the crux of the forecast process. In either case, the clouds and precipitation are caused by some type of lift, often called *ascent* by meteorologists. Stratiform rain is caused by slow, large-scale ascent of a very humid air mass. Convective showers and rain can be aided by this process, but depend much more on the presence of instability (very warm air underlying cold air). The ascent occurs in the form of very small pockets of rising motion, the size of a cumuliform cloud.

Sources of ascent

Slow, large-scale (synoptic-scale) ascent is one of the easiest problems for amateur forecasters, professionals, and numerical weather models to tackle. It occurs over such a large area that the air mass characteristics are sampled quite well by surface and radiosonde stations. In most cases, large-scale ascent is revealed in many different ways, including:

Upper-level forcing, often called "dynamics". This occurs when air is no longer in geostrophic balance, typically because of thermal contrasts in the air mass below. The air is forced to seek out sinking or rising motion to try to compensate for the lack of geostrophic balance. When the response is divergence in the upper atmosphere, which removes mass from the column and lowers surface pressures, air tends to rise to "fill the void". The famous "four-cell concept" of vertical motion and Q-vector divergence are all indicators of upper-level forcing.

Isentropic lift, which is most prominent when air parcels are travelling over rapidly varying air mass temperatures. The parcels must rise or sink in order to conserve their potential temperature.

Surface convergence, due to a clash in low-level wind direction or a low pressure area, causes ascent. Air converges and is forced to rise. This is closely related to orographic lift.

Orographic lift, where air is forced higher and higher as it travels along ascending terrain. This can occur on scales anywhere from ascent up a mountain to long runs of ascent measuring 1000 miles or more on the Great Plains. The longer runs of ascent are often referred to as upslope flow.

Air may also rise in a convectively unstable air mass to form showers and thunderstorms. This occurs when very cold air overlies warm, moist air in the low levels. Intense sunshine is not necessarily a requirement for convective instability. It can come from differential advection, for example when cold air overruns an area aloft, or from heating of an air mass by a warm body of water, such as Great Lakes lake-effect snowstorms.

Finding areas of ascent

Each of these mechanisms has indicators, patterns, and characteristics which show up on standard weather charts. For example, warm air advection (WAA) occurs where winds and pressure gradient are bringing in warmer air. This tends to imply the presence of isentropic lift. At the 500 mb level, cyclonic vorticity advection (CVA), also known as positive vorticity advection (PVA) in the Northern Hemisphere, is often associated with upper-level divergence, which implies upward motion.

Although there are model charts that depict

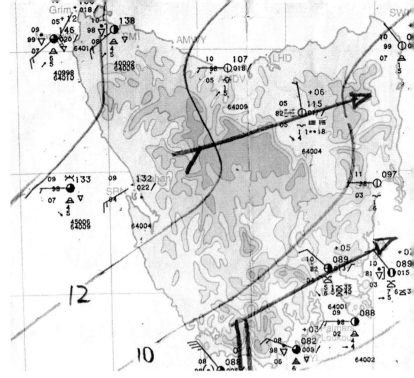

omega, which is the value of vertical velocity, these do not provide an easy, cookie-cutter solution. The omega fields tend to be noisy and difficult to scrutinize. Furthermore, they say nothing about why the lift or subsidence is occurring, its character, and what implications it has for the forecast area. A wise weather forecaster will look past these nondescript "head-lines" and open up the charts to get a full understanding of what is happening.

Unfortunately this is where the discussion must end. To jump into these issues in much greater detail, see titles like the author's *Weather Forecasting Handbook*, Jack William's *The Weather Book*, and Stanley David Gedzelman's 1980 *Science and Wonders of the Atmosphere* (if you can still get it!).

Other valuable resources

Though book knowledge will take you far, there are a few other things you need to know before you get started. Practice often, and get familiar with the UTC time zone!

Pencils and crayons. If you are planning to dig deep into chart analysis, a good supply of HB and 2H pencils, as well as crayons, are a great idea. Keep a box full of them near your analysis table. A good-quality drafting eraser will help scrub out mistakes. Various colored highlighters have all kinds of uses, including the ability to make working marks in your atlases without creating too much of a mess. Finally, a divider is handy for translating map distances on map scales. Most of these items can be purchased at art and hobby shops, office supply stores, and architectural supply companies.

Atlases. Topography and geography plays into all aspects of meteorology, from orographic lift to storm reports from small towns. National Weather Service analysis desks are characterized by their excellent assortment of beat-up atlases. Likewise, keep a few titles near your desk. For the United States, the *Michelin Road Atlas and Travel Planner* edges out ahead of the standard Rand

Above: Analysis chart for Tasmania. *(Australian Bureau of Meteorology)*

Section title photo: Forecasters working on a severe thunderstorm problem at the Fort Worth WSFO in Texas. *(Tim Vasquez)*

McNally and AAA atlases since it offers full map coverage at a constant scale and avoids boxing in all the pages to specific states. This makes it very easy to track weather systems and visualize their size as they progress across state and national borders. For worldwide topographic views the *Planet Earth Macmillan World Atlas* maps out the world at two sets of equivalent scales. Try not to be edgy about drawing in your atlases and bending the spine to get a better look; after all they're there to be used rather than to collect dust in your bookshelf, and used replacements are cheap.

Coordinated Universal Time (UTC). The UTC time zone is used religiously in weather forecasting, and you will not get far without knowing how to convert your time zone to UTC and back. UTC time is simply the time in London, England, not counting for their daylight saving time. See Table 1 for exact conversions.

Practice, practice, practice! Skills only come with experience, and experience cannot be built if you are taking a casual, uninvolved approach to the weather. Supplement your experience by building a good weather library and using it.

Table 1. Time zone conversions. The "code" is the military designator, typically expressed as the phonetic word for each letter (Alpha, Bravo, etc). For further information see <*www.timeanddate.com*>.

When standard time is in effect (no local daylight saving or summer time)

Code	If your time zone is	To convert from UTC to local	Examples
Y	International Date Line West (IDLW)	subtract 12 hours	(none)
X	Samoa Standard Time (SST)	subtract 11 hours	Apia, Niue, Midway, Pago Pago
W	Hawaii Standard Time (HST)	subtract 10 hours	Honolulu, Hilo, Tahiti
V	Alaska Standard Time (AKST)	subtract 9 hours	Anchorage, Fairbanks, Barrow
U	Pacific Standard Time (PST)	subtract 8 hours	Los Angeles, Seattle, Vancouver
T	Mountain Standard Time (MST)	subtract 7 hours	Phoenix, Salt Lake City, Denver, Calgary
S	Central Standard Time (CST)	subtract 6 hours	Dallas, Chicago, Minneapolis, Winnipeg
R	Eastern Standard Time (EST)	subtract 5 hour	New York City, Boston, Toronto
Q	Atlantic Standard (AST), Western Brazil (WST)	subtract 4 hours	Halifax, Moncton, Bermuda, Thule
-	Newfoundland Standard Time (NST)	subtract 3.5 hours	Stephenville, St. Johns
P	Brazil Time (BRT), Western Greenland Time (WGT)	subtract 3 hours	Nuuk, Buenos Aires, Rio de Janeiro
O	Fernando de Noronha Time (FNT)	subtract 2 hours	Noronha
N	Azores Time (AZOT), Eastern Greenland (EGT)	subtract 1 hour	Scoresbysund, Cape Verde
Z	Greenwich Mean Time (GMT)	no change	London, Glasgow, Dublin, Lisbon, Dakar
A	Central European Time (CET), W. Africa Time (WAT)	add 1 hour	Paris, Frankfurt, Stockholm
B	E. European Time (EET), Central Africa Time (CAT)	add 2 hours	Helsinki, Kiev, Sofia, Athens
C	Arabian Standard Time (AST), E. Africa Time (EAT)	add 3 hours	Moscow, Baghdad, Aden, Nairobi
D	Gulf Standard Time (GST), Russia Zone 4 (ZP4)	add 4 hours	Samara, Dubai, Muscat
E	Pakistan Time (PKT), Russia Zone 5 (ZP5)	add 5 hours	Yekaterinburg
-	India Standard Time (IST)	add 5.5 hours	Calcutta, New Delhi, Bombay
F	Bangladesh Time (BDT), Russia Zone 6 (ZP6)	add 6 hours	Omsk, Dhaka
G	Indochina Time (ICT)	add 7 hours	Krasnoyarsk, Bangkok, Jakarta, Hanoi
H	Aust. Western Std. Time (AWST), China Std. (CST)	add 8 hours	Perth (WA), Manila, Taipei, Beijing
I	Japan Standard Time (JST), Korea Standard (KST)	add 9 hours	Tokyo, Seoul, Yakutsk
-	Australia Central Standard Time (ACST)	add 9.5 hours	Darwin, SA/NT
K	Australia Eastern Standard Time (AEST)	add 10 hours	Sydney, QL/ACT/NSW/VIC/TAS
L	Magadan Time (MAGT), New Caledonia (NCT)	add 11 hours	Magadan, Noumea, Ponape
M	New Zealand Standard Time (NZST)	add 12 hours	Kamchatka, Auckland, Kwajalein, Fiji
-	Tonga Time (TOT), Phoenix Isl. Time (PHOT)	add 13 hours	Enderbury, Tongatapu
-	Line Island Time (LINT)	add 14 hours	Kiritimati

When local daylight saving or summer time is in effect

Code	If your time zone is	To convert from UTC to local	Examples
U	Alaska Daylight Time (AKDT)	subtract 8 hours	Anchorage, Fairbanks, Barrow
T	Pacific Daylight Time (PDT)	subtract 7 hours	Los Angeles, San Francisco, Vancouver
S	Mountain Daylight Time (MDT)	subtract 6 hours	Denver, Salt Lake City, Calgary
R	Central Daylight Time (CDT)	subtract 5 hours	Dallas, Chicago, Minneapolis, Winnipeg
Q	Eastern Daylight Time (EDT)	subtract 4 hours	New York City, Boston, Toronto
P	Atlantic Daylight Time (ADT)	subtract 3 hours	Halifax, Moncton, Bermuda
A	British/Irish Summer Time (BST)	add 1 hour	London, Belfast, Glasgow
B	Central European Summer Time (CEST)	add 2 hours	Frankfurt, Paris, Oslo, Zurich, Warsaw
C	Eastern European Summer Time (EEST)	add 3 hours	Helsinki, Athens, Sofia, Kiev
D	Moscow Summer Time (MSD)	add 4 hours	Moscow
K	Japan Daylight Time (JDT)	add 10 hours	Tokyo
-	Australian Central Daylight Time (ACDT)	add 10.5 hours	Adelaide, SA
L	Australian Eastern Daylight Time (AEDT)	add 11 hours	Sydney, ACT/NSW/VIC/TAS

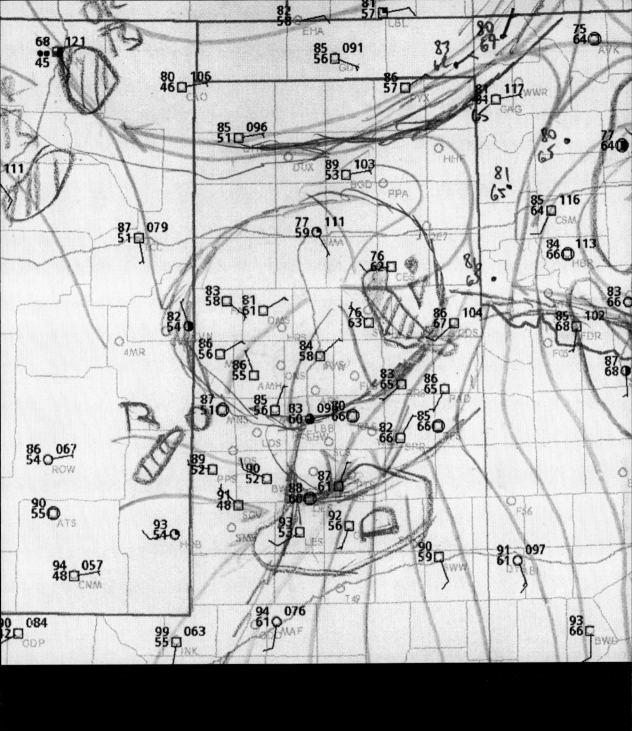

CHAPTER 1

ANALYSIS CHARTS

Observational Charts

A weather product is often used merely to obtain the basic information that it shows. What is the temperature? Is it raining over Pittsburgh? Is there a threat of clouds for tonight's star party? Will the flight crew have a good tailwind?

In a more organized forecast setting, meteorologists are not as interested in exact numbers as in being able to visualize processes in the atmosphere and look for clues. Therefore this book attempts to explain not only how to interpret maps but also to find the subtle signals, patterns, and trends that may have a much greater impact on the forecast. This is the science and art of meteorological diagnosis. Developing the talent for diagnosis requires a little philosophical insight and the willing participation of the forecaster.

The chart analysis process

Some excellent papers have been written on the philosophy behind using weather charts. Two of them are, "The Role of Diagnosis in Weather Forecasting" (1986) by Charles Doswell III and Robert Maddox; and "The Human Element in Weather Forecasting" (1986) by Charles Doswell III, both online at <www.cimms.ou.edu/~doswell>. In these essays, Doswell and Maddox identified two separate steps that must occur before a prognosis (forecast) can be made: analysis and diagnosis.

The first step, analysis, is the identification of ingredients in the atmosphere. It is the process of drawing lines and identifying fronts, lows, highs, wind shift lines, jets, and air masses. A computer may assume a lot of the analysis procedures. Following this, a process called diagnosis is performed. This is composed of thought, ideas, and conclusions. It is the process of visualizing the completed analysis and relating it to other products. Diagnosis is also the precursor to a coherent, robust forecast.

Unfortunately in today's computer-driven age, some forecasters complete the analysis but fail to perform any diagnosis. Quick glances at the maps and reliance on computer-drawn isopleths do not constitute diagnosis. As a result, the misguided forecast process inevitably leans toward very heavy emphasis on numerical models. The balance in the forecast process is spelled out surprisingly clear in forecast discussions, case studies, and even technical papers.

There are countless situations where important ingredients are resolved not through the models but only through careful analysis and diagnosis. One case in point is a jet max moving out of New Mexico on 3 May 1999, linked to a deadly tornado outbreak near Oklahoma City, and initially detected only by the Tucumcari NM wind profiler. In this case, meteorologists detected the clue and worked it into the forecast.

A debate has sometimes emerged among some of the best operational forecasters: is it absolutely necessary to put pencil to paper and "hand-analyze" the chart? The consensus is generally "yes". Either way, there is nothing to be lost by forcing yourself to scrutinize the data in better detail and find a relationships in the patterns.

Even so, hand analysis only works when the forecaster makes an effort to think about the data. Drawing lines with a closed or distracted mind accomplishes nothing more than drawing lines. Make an effort to visualize the meaning of the data as you put pencil to paper. Glance at the station plots as you go, picturing the weather and contrasting it to conditions you see nearby. The hand analysis is not a picture to be drawn; it is a canvas for thought.

Frontal placement

Fronts are one of the most basic components of a weather chart, as they highlight *baroclinic* zones, where temperature advection is taking place and atmospheric energy is at work. A front is always drawn with its barbs or pips facing the direction of movement. Table 2 details the styles used for depicting a front.

Fronts are always placed on the *warm side of a temperature gradient*. This makes perfect logical sense, as when we picture a cold air mass moving into a region, the front occurs when the temperature first begins falling, not when the temperature drop is complete. When warm air is invading, it is not quite as intuitive, but the most basic weather books demonstrate that the warm front passes after the temperature has finished climbing (not counting the effects of diurnal heating, of course).

Drylines are sharp boundaries between a moist tropical air mass and a very dry continental airmass, usually originating from higher terrain. They are common in the southern Great Plains of the United States but may also be found in India, the Sahel, and Australia. The dryline is always located on the *moist side of the dewpoint gradient*. The barbs point toward the direction of the dryline's movement.

Finally, troughs and other features can be sketched in. However it should be noted that it is important to spend more time refining your position of the feature than attempting to categorize it. The feature can be revisited later in the diagnosis process and may be much more meaningful then.

The symbols for fronts and boundaries are shown in Table 2. Throughout this book, various styles and colors for other parameters are recommended. Most of these assume you are marking on white charts. For black (on-screen) graphics, these colors are all still valid, except that white and black are reversed.

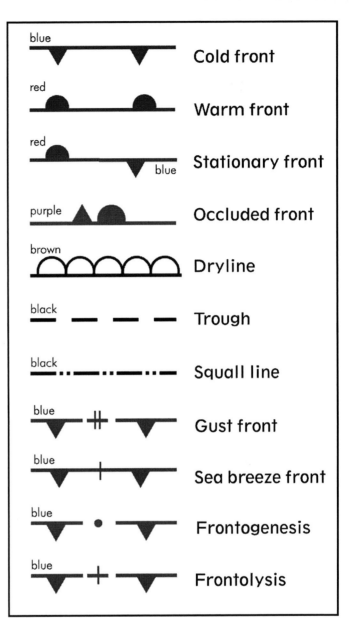

Table 2. Conventions for drawing fronts and boundaries. The scheme for a gust front and sea breeze front is adopted from Young and Fritsch (1989).

Surface chart

The surface chart is the backbone of weather forecasting and is the most familiar map to anyone who has dabbled in meteorology. The first bona fide weather map of simultaneous conditions across a region was a chart drawn in 1815 by German scientist Heinrich W. Brandes.

The biggest advantage of the surface chart is that data is available quite frequently: as often as six hours in remote regions and as much as every 20 minutes in North America. A less obvious but extremely important use of the chart is to search for imbalances that reflect processes occurring aloft, which may not be reflected by the coarse radiosonde data that is only available every twelve hours. For example, pressure fall centers or very gusty winds may be closely linked to areas of strong upper-level forcing moving with the upper-level flow.

Although winds closely follow the pressure (height) lines on upper-level charts, this is not necessarily the case on the surface chart. The winds usually turn more directly toward low pressure. This is because the effect of friction is much more pronounced, which diminishes the Coriolis effect and allows air to move more directly toward low pressure. The effect is not so strong on ocean surfaces due to weaker friction.

Fronts are frequently found on the surface chart. The primary indicator of a frontal boundary is the temperature field. In mountainous areas it can be helpful to analyze using theta, or potential temperature, to normalize the effects of elevation.

A recommended sequence for analyzing the surface chart is to look for obvious fronts and boundaries and sketch them in lightly. Then sketch in isobars lightly with the pencil. Use the isobars and other indicators, and even other tools such as satellite and radar, to further refine front and boundary positions. Then "harden in" the final position of the front. Following this, "harden in" the isobars, forcing them to kink along the front. While this process is underway, never forget to spend time visualizing what is seen on the chart, feeling the data as you go. This will help make your diagnosis complete.

It must be emphasized that the annotations drawn on the chart are not nearly as important as the process of diagnosing those actual features and understanding their impact on the forecast. A hand analysis is a canvas for thought and meditation on the forecast problem, with the goal of mentally absorbing and assimilating the maximum amount of information offered by the chart.

a closer look

Surface plots are explained in Table 1A in the Appendix.

Observations of U.S. surface weather have been made mostly by automated sensors since the mid-1990s. Human weather observations are still widespread outside North America and Europe.

The vast majority of surface data used by meteorologists flows through data circuits in either METAR or SYNOP format. METAR is used mostly in North America and Europe, with SYNOP elsewhere. Some specialized networks use databases or proprietary tabular formats, such as Australia's detailed surface data.

Fields are drawn as follows:
- Isobars are drawn in solid black every 4 mb or 0.05 in Hg.
- Isotherms (optional) are drawn every 2 C° or 5 F°.
- Add other fields as needed.

Features are drawn as follows:
- High and low pressure centers are marked using, respectively, a large blue "H" or a large red "L". The center pressure value may be labelled below the H or L, typically in tens and units of a millibar.
- Fronts are drawn in blue and red using standard notation.
- Trough axes are drawn as a thick black dashed line.
- Drylines are drawn as a thick brown line with hollow pips connected to one another, facing toward the moisture.

websites

www.hpc.ncep.noaa.gov/html/sfc2body.html
www.rap.ucar.edu/weather/surface
weather.cod.edu/analysis
www.hko.gov.hk/wxinfo/currwx/wxcht.htm
www.spc.noaa.gov/exper/mesoanalysis
weather.unisys.com/surface

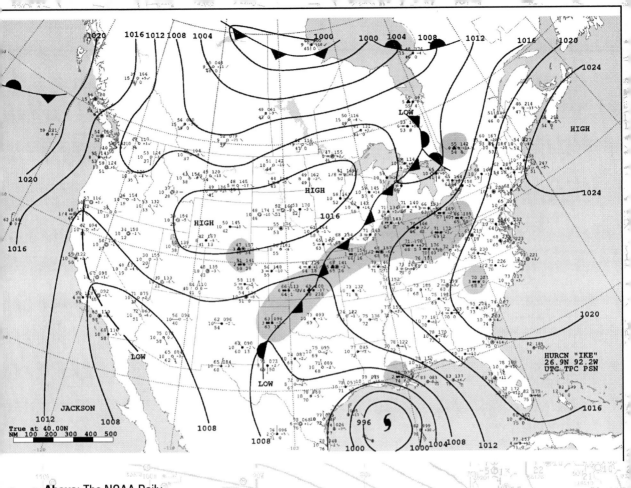

Above: The NOAA Daily
Weather Map series, pro-
duced almost every day
since 1871, constitutes
the "official" United States
surface chart. It is now
produced by the NCEP
Hydrometeorological Pre-
diction Center. Preliminary
versions are created every
three hours on the HPC
website. *(NOAA)*

Right: Surface chart
produced under Windows
using the Digital Atmo-
sphere package. Interest-
ingly this particular chart
was configured to use the
legacy NMC fonts, which
were developed in 1969
by programmer Gloria Dent
and were widely used in
NWS fax charts during the
1970s and 1980s.

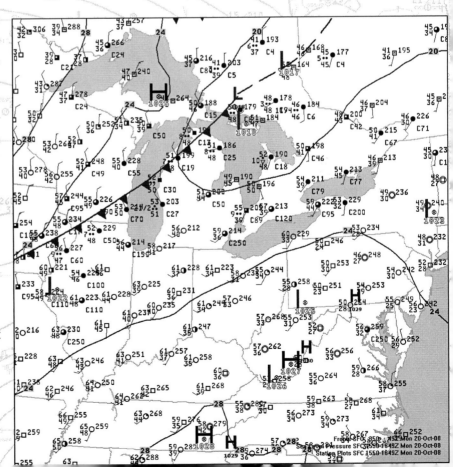

850 mb chart

The 850 mb level is located at about 5000 ft MSL, assuming a standard atmosphere, however it can vary by several hundred feet or more. The key phrase is MSL: mean sea level. This means that along flat land near sea level, the 850 mb level will be about 5000 ft above the ground. At Salt Lake City, 4300 ft elevation, the 850 mb level is only about 700 feet above the ground! And in the Rocky Mountains, the 850 mb level is far underground. It can be said that the 850 mb level may be found in one of three places: underground, within the planetary boundary layer, or in the free atmosphere.

The planetary boundary layer (PBL) is that part of the troposphere in contact with the ground. A typical depth is about 2000 ft. The PBL is strongly influenced by the earth's surface through heating, drag, and evapotranspiration. In reality, the top of the PBL can range from the Earth's surface on a clear, cold night to 10,000 ft AGL or more given a windy, unstable, uncapped air mass with very strong heating.

Therefore depending on the surface elevation, the time of day, the season, and the ongoing weather regime, the 850 mb level, if not underground, will be within the PBL or within the free atmosphere. It is useful to a sense of which layer the radiosonde is representing and get familiar with all of the radiosonde station elevations in your region to see how high the 850 mb level usually is at each site. It is also important to bear in mind that the PBL and free atmosphere can seem to merge, especially during the summer in weak wind patterns, and drawing a distinction may not be important.

When it comes time to put the pencil to the maps, the most popular use for the 850 mb chart is to locate frontal systems, particularly when their positions are not clear on the surface chart. Remember that frontal surfaces slope up and into cold air, therefore the 850 mb front will almost always be found poleward of the surface front. There are some exceptions, particularly on the Plains, where cold fronts will move much faster aloft than near the ground, and this may result in a cold front being placed equatorward of the surface front.

During the spring months, the 850 mb chart helps paint out the configuration of the low-level jet (LLJ). This feature is common before severe weather outbreaks and typically stretches from the Gulf of Mexico coastline into the central United States. The depth of tropical air masses can be assessed by determining the dewpoint at 850 mb and relating that to surface dewpoints.

a closer look

The 850 mb chart is usually near 5000 ft MSL. It is useful for finding the structure of frontal systems aloft, for assessing the depth of tropical and polar air masses alike, and for finding low-level jets. It intersects with the typical level of cumulus and stratocumulus clouds.

Upper air plots are explained in Table 1B in the Appendix.

Fields are drawn as follows:
- Height contours are drawn as solid black lines every 30 m (3 dam) using 150 dam as a base value (147, 150, 153, etc). Labels are in dam.
- Isotherms are drawn as dashed red lines every 5 C°, however 2 C° and 4 C° intervals are common.
- The type of moisture analysis (optional) is contingent on the type of weather regime:
  -- In stratiform situations, a relative humidity parameter is recommended. Shade all areas in green with dewpoint depressions below 5 C° (this is roughly a 75% relative humidity).
  -- In convective situations, use an absolute humidity parameter. Figure the dewpoint for each station and draw isodrosotherms.

Features are drawn as follows:
- Low level jet axes (narrow bands of strong winds exceeding 40 kt) should be drawn as a thick red arrow.
- Moisture axes may be drawn as thick, wavy green lines. Pencil is best.
- Fronts should be drawn in standard blue and red colors. Since this is an upper-air chart, do not shade the barbs and pips.

websites

weather.noaa.gov/fax/nwsfax.html
www.weatheroffice.gc.ca/analysis
www.rap.ucar.edu/weather/upper
weather.cod.edu/analysis
www.spc.noaa.gov/exper/mesoanalysis
weather.unisys.com/upper_air

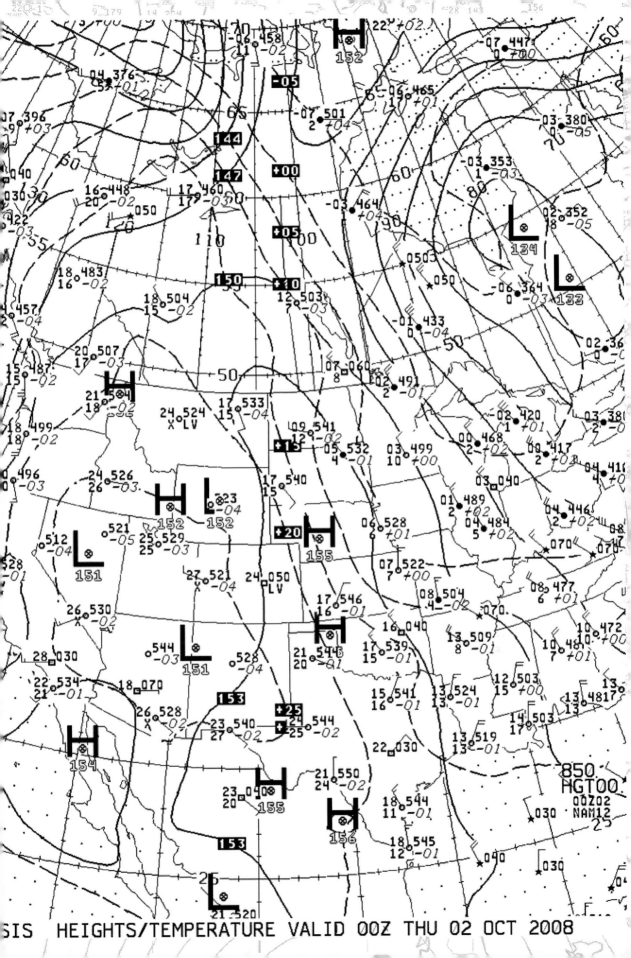

SIS HEIGHTS/TEMPERATURE VALID 00Z THU 02 OCT 2008

700 mb chart

The 700 mb level is roughly at 10,000 ft MSL. This is considered to be somewhere between the lower and middle troposphere. At higher radiosonde stations such as Denver and Salt Lake City, the 700 mb level is only about 5000 ft above the ground and may be within part of the planetary boundary layer during the warm season (see section on the 850 mb level). This is especially true of the higher mountainous regions.

Weather systems at 700 mb typically take on an open, broad look compared to the patterns at lower levels. The fronts are usually found further poleward compared to the 850 mb and surface charts, owing to the slope of fronts up and into the cold air. This relationship can help place the low-level front when the 700 mb chart shows a thermal gradient and the 850 mb and surface charts don't have a clear position.

The relationship of 700 mb flow to a surface cold front is a rule of thumb indicating what type of front is present. A katafront is associated with a significant component from the cold to warm air mass; an anafront with a neutral component or one from the warm to the cold air mass at 700 mb.

Considering that the vast majority of cloud cover in a developing frontal weather system occurs in the 5,000 to 15,000 ft MSL range, humidity is a favorite quantity for measuring the extent of synoptic-scale lift and moisture. Relative humidity values of 70% or dewpoint depressions of 5 C° or less at 700 mb are considered synonymous with overcast cloud cover. The relative humidity value is usually expressed on model output panels, while the dewpoint depression is shown on upper-air plots. When areas exceeding these threshold values are shaded in green, a definition of the area and shape of strongest upper-level forcing emerges. A "wrapped" structure may even be seen on the charts, matching quite well with the cloud bands observed on satellite imagery.

In springtime thunderstorm situations, the 700 mb level is usually within the heart of the elevated mixed layer (EML), a broad area of warm, dry air originating from the southwestern United States that is lofted eastward into the central United States. The air at this level is often warmer than that below it, which provides an inversion, or "cap", that either prevents thunderstorms altogether or suppresses them until afternoon when heating and instability are maximized. Therefore the 700 mb isotherm pattern can help define the coverage and strength of the EML.

a closer look

The 700 mb chart is usually at about 10,000 ft MSL. It is used to find mid-level wind cores, short waves, capping inversions, and areas of mid-level clouds and moisture. It intersects with the level where mid-level clouds such as altocumulus and altostratus are found.

Upper air plots are explained in Table 1B in the Appendix.

Surface lows tend to move at about 70% of the 700 mb wind speed above the low.

Fields are drawn as follows:
- Height contours are drawn as solid black lines every 30 meters (3 dam) using 300 dam as a base value (297, 300, 303, etc). Labels are in dam.
- Isotherms are dashed red lines every 5 C°. Intervals of 2 C° or 4 C° are common.

Features are drawn as follows:
- Areas of significant moisture (dewpoint depression of 5 C° or less) may be shaded in green.
- Jets axes are represented by a thick red arrow.
- Fronts should be drawn in standard blue and red colors. Since this is an upper-air chart, do not shade the barbs and pips.
- Short wave trough axes are drawn as thick black straight lines.
- Short wave ridge axes are drawn as thick black zig-zag lines.
- Col. Robert Miller's Severe Weather Analysis Notes suggests coloring jets and moisture in brown for this level to signify that they apply to 700 mb.

websites

weather.noaa.gov/fax/nwsfax.html
www.weatheroffice.gc.ca/analysis
www.rap.ucar.edu/weather/upper
weather.cod.edu/analysis
www.spc.noaa.gov/exper/mesoanalysis
weather.unisys.com/upper_air

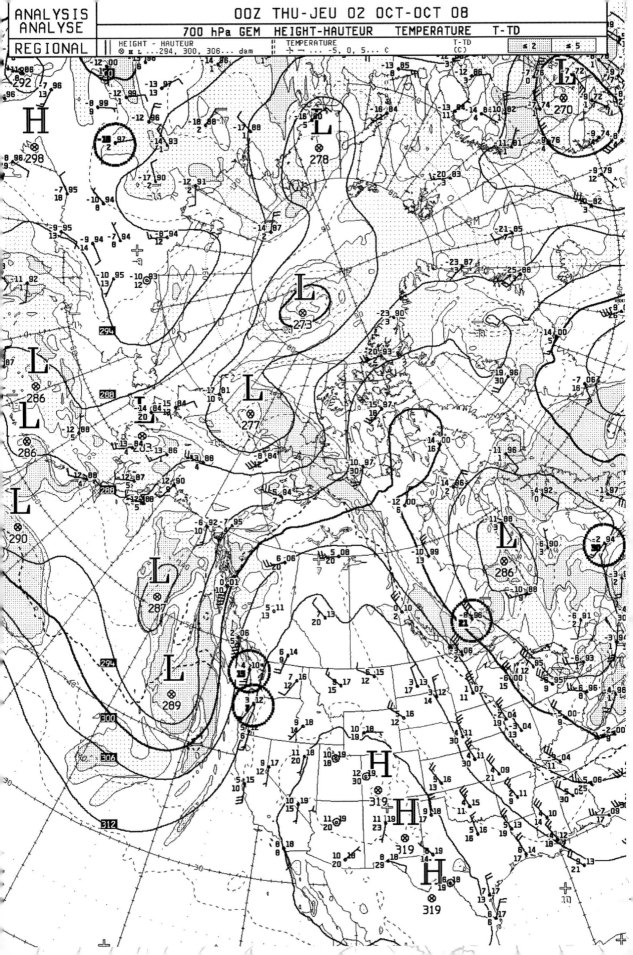

500 mb chart

Except perhaps in polar regions, the 500 mb level is considered the middle of the troposphere. It exists at a height of about 18,000 ft MSL, and so is nearly always within the free atmosphere and not part of a planetary boundary layer.

At this level, forecasters see an excellent mix of small-scale and large-scale systems. At the large scale, the upper-level jet pattern begins emerging, painting out the regions of strongest baroclinicity (energy available to the atmosphere). Superimposed on this are a series of large troughs and ridges. The troughs correspond to very cold air masses, while the ridges exist above areas of warmth.

At even smaller scales we find the notorious short waves. These are small-scale troughs and ridges embedded in the large-scale flow which are just barely resolved in the existing radiosonde network. Short waves are reflections of important thermal perturbations in the air mass beneath, whose influence easily translates to higher levels.

These perturbations, if a strong thermal gradient is present in the lower troposphere, often go on to amplify the short wave which in turn deepens the short wave, which in turn imparts more energy to the system beneath. This is a chain reaction called "self development" and is broken only when the system finally occludes and washes out the thermal gradient. The short wave trough, by this time, has often deepened into an upper-level low.

Short waves are located using either the absolute vorticity field or by looking for wind shifts. On standard upper-air analyses, vorticity overlays are sometimes not available. The vorticity fields are largely a product of numerical weather prediction output, and the short waves tend to lie along elongations of vorticity axes, particularly those that cross the flow in a perpendicular orientation. The short wave will tend to be located along a boundary oriented in a perpendicular fashion across the 500 mb flow which separates two stations with a sharp wind shift. A cyclonic wind shift defines a short wave trough, while an anticyclonic wind shift indicates the presence of a short wave ridge.

Finally, it is crucial to make that distinction between a "short wave" and a "short wave trough", because the term short wave applies to both. Also short waves can sometimes be resolved at lower levels, such at the 700 mb level. Short wave troughs stack downward toward the warmer air (upward toward colder air).

a closer look

The 500 mb level is usually at a height of about 18,000 ft MSL. It is used for finding mid-level jet, for seeing the general overview of the tropospheric pattern, and is the primary chart for finding short waves. It is near the level of non-divergence (the center of mass in the troposphere).

Upper air plots are explained in Table 1B in the Appendix.

Surface lows tend to move at about 50% of the 500 mb wind speed above the low.

Fields are drawn as follows:
- Height contours are drawn as solid black lines every 60 m (6 dam) using 570 dam as a base value (564, 570, 576, etc). Labels are in dam.
- Absolute vorticity contours are machine-produced and are typically drawn as dashed black lines every $2*10\text{-}5$ sec.
- Isotherms (optional) are dashed red lines every 5 C°. Intervals of 2 C° or 4 C° are common.

Features are drawn as follows:
- High and low height centers are plotted as a large black "H" or "L", with the decameter value below it.
- Areas of PVA are shaded red.
- Areas of NVA are shaded blue.
- Short wave trough axes are drawn as thick black straight lines.
- Short wave ridge axes are drawn as thick black zig-zag lines.
- Jet axes are drawn as a thick red line.
- Col. Robert Miller's Severe Weather Analysis Notes suggests a blue color for any jet depiction to signify that the markings are for 500 mb.

websites

weather.noaa.gov/fax/nwsfax.html
www.weatheroffice.gc.ca/analysis
www.rap.ucar.edu/weather/upper
weather.cod.edu/analysis
www.spc.noaa.gov/exper/mesoanalysis
weather.unisys.com/upper_air

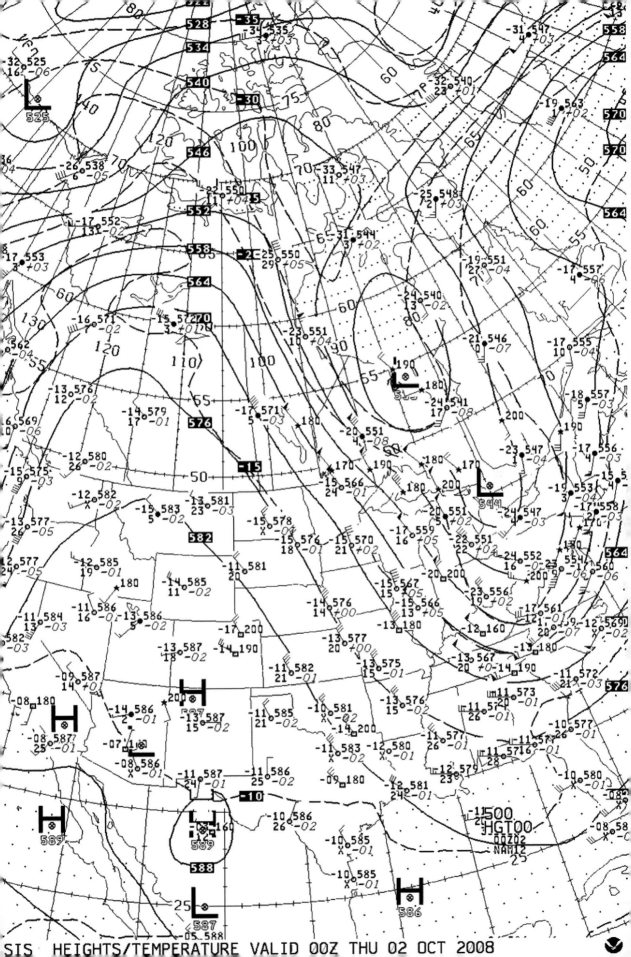

SIS HEIGHTS/TEMPERATURE VALID 00Z THU 02 OCT 2008

300/250/200 mb chart

The practicing meteorologist always wants to have a glance at the upper tropospheric conditions, as the polar front jet lies in its topmost portions. Unfortunately picking a level is complicated, because the top of the troposphere (the tropopause) in temperate latitudes ranges in height from 30,000 ft during the winter to 45,000 ft during the summer.

Therefore, given the standard levels of 300 mb (30,000 ft MSL), 250 mb (34,000 ft MSL), and 200 mb (39,000 ft MSL), it is necessary to choose different charts depending on the season to find most representative "upper tropospheric chart". During the winter, 300 mb is used. In the transition seasons of autumn and spring, the 250 mb chart is preferred. In the summer, the 200 mb is selected. The others are discarded as they are either too low or tap into the stratosphere where the polar jet is rapidly weakened with height. Though there is little use for stratospheric charts, some studies have been published on stratospheric warm sinks and cold domes, which have ties to areas of upper divergence and convergence, respectively.

Overall, the axis of strongest winds paints out the jet stream. This pattern is by far the highlight of the upper tropospheric chart, and defines the weather regime that is in place. Long waves are formed by the very broad troughs and ridges that ring the hemisphere. The long wave troughs are caused by large masses of cold air, and the ridges by warm air.

Jet maxes, sometimes called jet streaks particularly when referring to smaller scales, are very important features. The winds are frequently out of balance around them, resulting in strong vertical motions. A careful isotach field that provides the correct shape of the jet streak can be very helpful in inferring areas of vertical motion.

A conceptual model exists which suggests the type of vertical motion that may exist around a jet max. It is usually referred to as the "4-cell jet max concept". The coordinate system specifies that "left" is the poleward direction, "right" is equatorward, "rear" is upstream, and "front" is downstream. Upward motion should be found in the left front quadrant (LFQ) and right rear quadrant (RRQ), with downward motion in the right front quadrant (RFQ) and left rear quadrant (LRQ). In cyclonic flow the left quadrants are enhanced with the right quadrants nullified, with the opposite true in anticyclonic flow. This concept is highly subjective and is subject to assumptions, but demonstrates an excellent use for the upper tropospheric chart.

a closer look

The 300 mb chart is usually at a height of about 30,000 ft. The 250 mb chart is usually at a height of about 34,000 ft. The 200 mb chart is usually at a height of about 39,000 ft. It is used primarily to establish the pattern of upper-level jets, to locate areas of shear, and to assess the nature of the hemispheric circulation and identify any blocking patterns. All levels intersect with the height regime where cirriform clouds are found.

Fields are drawn as follows:
- Height contours are drawn as solid black lines every 120 m (12 dam) using 900 dam as a base value (888, 900, 912, etc). Labels are in dam.
- Isotachs are drawn as purple lines every 20 kt starting at 30 kt as a base (30, 50, 70, 90, etc).
- Isotherms (optional) are dashed red lines every 5 C°, but may clutter the chart.

Features are drawn as follows:
- High and low height centers are plotted as a large black "H" or "L", with the decameter value below it.
- Jet axes. The axis of jets should be drawn as a heavy red arrow. Col. Robert Miller's Severe Weather Analysis Notes suggests a purple color for this level.
- Jet cores. Draw a red ellipse to mark the jet core.

Isotach bands can be shaded using colored pencils.
- One possible shading spectrum is green to yellow to red.
- Another spectrum is green to red to blue to purple to yellow. This scheme is used by Environment Canada.

websites

weather.noaa.gov/fax/nwsfax.html
www.weatheroffice.gc.ca/analysis
www.rap.ucar.edu/weather/upper
weather.cod.edu/analysis
www.spc.noaa.gov/exper/mesoanalysis
weather.unisys.com/upper_air

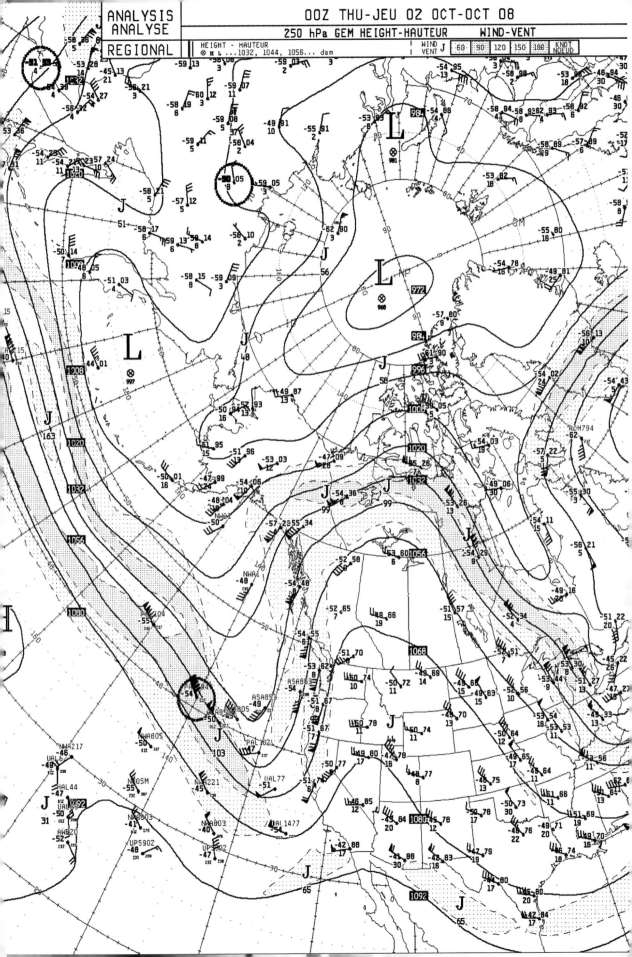

Thickness chart

Thickness is a direct measurement of the vertical distance between one pressure surface and another. By convention it is expressed in whole meters, or more commonly, as dekameters (tens of meters), abbreviated as "dam". Contours are usually drawn every 6 dam. Thickness is useful to forecasters because it provides an indication of the average temperature within the layer being sampled. Overall, high thickness values are always associated with warm air, while low thickness values are associated with cold air. The 540 dam line has often served as a rough transition line between snow and rain.

To be more exact, thickness measures the average virtual temperature, which is only slightly different from actual temperature by no more than one or two Celsius degrees. At any given virtual temperature (i.e. any given thickness) an air mass will be slightly cooler if it is humid and slightly warmer if dry.

The most common layer found in operational meteorology is the layer between 1000 and 500 mb, which yields the 1000-500 mb thickness. This layer, which occupies the area roughly between sea level and 18,000 ft MSL, represents the bottom half of the troposphere where the majority of air mass contrasts exist. During extremely cold events that involve much more shallow air masses, the 1000-700 mb or even the 1000-850 mb thickness may be used to better define fronts and air masses. However such charts are extremely difficult to find on the Internet, and as a result are only available to those with weather display software.

Thickness plots are almost always displayed together with isobars, which connect lines of equal pressure. This provides an accurate relationship of the pressure gradient (and wind) to the thermal contrasts and air masses that exist.

Thickness charts provide one of the most reliable ways to assess thermal advection. Advection occurs where wind is blowing colder or warmer thicknesses into a given location. Where "boxes" are painted out by the isobars and thickness lines, it implies that either warm or cold advection is taking place.

Specific thickness lines have been used for decades to highlight rain-snow transition areas. The most common transition line is the 540 dam line (usually ±4 dam) on the 1000-500 mb thickness chart, which correlates to the rough location of rain-snow transition at sea level. The corresponding line on the 1000-700 mb chart is 284 dam and on the 1000-850 mb chart it is 130 dam.

a closer look

Thickness is an indicator of average temperature through a given layer of the atmosphere, usually from 1000 to 500 mb. Thickness lines can be thought of as isotherms for the entire layer. Thickness charts are used to locate and delineate fronts, air masses, and areas of thermal advection.

The 540 dam thickness line has been used for decades as a basic rule of thumb for locating the transition line between rain and snow. This does not differentiate ice pellets and freezing rain against other forms of precipitation.

Values: Low thickness corresponds to cold air. High thickness implies warm air.

Advection: Warm advection is associated with large-scale ascent, clouds, and rain. Cold advection is associated with subsidence and clear skies.

Fronts: Cold fronts tend to be associated with cyclonically curved thickness lines; warm fronts are associated with anticyclonically curved thickness lines.

Bands of thickness lines are thermal gradients which tend to separate different air masses. A surface front usually exists on the warm edge of bands of thickness lines (thermal gradients).

A surface low riding along the warm side of a thickness gradient is developing or mature. When it recedes within or poleward of the thickness gradient, it is occluding.

websites

weather.noaa.gov/fax/nwsfax.html
weather.cod.edu/forecast
weather.unisys.com/eta/pres.html
www.uga.edu/atsc/wx/ncep_loops.htm
www.weatheroffice.gc.ca/model_forecast

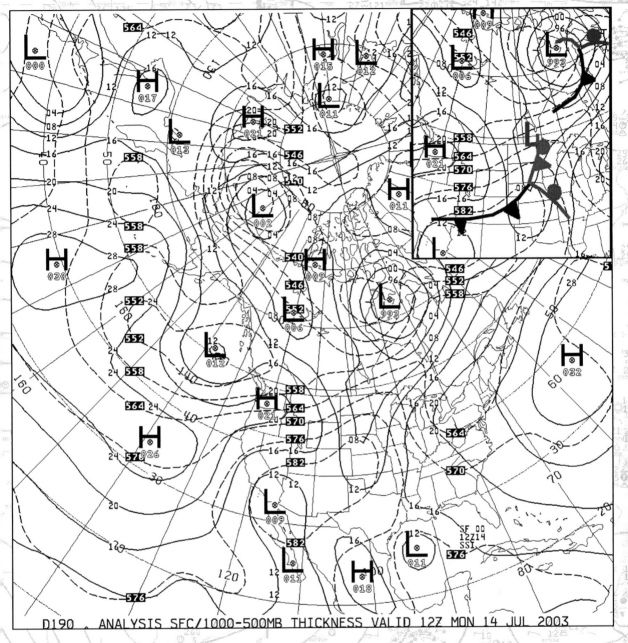

D190 . ANALYSIS SFC/1000-500MB THICKNESS VALID 12Z MON 14 JUL 2003

Above: Thickness chart for the 1000-500 mb layer. Note the "boxes" formed by the juxtaposition of the thickness and pressure lines, which imply thermal advection. Strong cold advection exists over Montana and Wyoming, while strong warm advection exists over Wisconsin and Minnesota. A frontal system (inset) is suggested by the patterns. The pressure-thickness chart is an excellent tool that can be used to infer frontal systems from model output, even when surface data is not available. *(NOAA/NCEP)*

Isentropic analysis

Isentropic analysis works on the understanding that a parcel of air does not move in a pure horizontal fashion but will cling to its own isentropic (potential temperature, i.e. theta) surface. This is true if there are no special thermal processes taking place involving addition or removal of heat, i.e. heating, evaporation, and condensation. As a result, vertical motion can be easily assessed simply by looking at the slope of the isentropic surfaces and the direction of the winds along the surface.

Surprisingly, isentropic analysis was widely used after WWII, but was largely abandoned with the first surge of model data in the 1960s. It wasn't until the early 1990s that the availability of instantaneous isentropic diagnosis on personal computers helped revive the technique. It is now in common use at most forecasting offices.

The basis of isentropic analysis is the rule that theta surfaces bend upward above a cold air mass (forming a sort of "cold dome" of theta surfaces) and dip downward over a warm air mass. This implies that when a mid-level parcel moves from a region of low-level warm air to another region with low-level cold air, it must rise to follow the upward bend in the isentropic surface. This is equivalent to the concept of "overrunning" ascent that occurs along warm fronts. Likewise, a movement of air from a cold to a warm region suggests isentropic descent, and this explains the rapid clearing that often takes place along cold fronts, particularly in their wake.

With the isentropic analysis chart, isobars will be displayed. These can be thought of as indicating the height of the theta field. For example, if we are looking at a 300 K isentropic level and a isopleth marked "700" crosses Vermont, we can conclude that the 300 K surface is at the 700 mb level in Vermont, which is about 10,000 ft above the ground. Therefore pressure contours with low values indicate high heights, and high values indicate low heights.

The second important component of the isentropic chart is wind barbs or wind vectors, which are valid for that isentropic level. So if a wind plot with southwest winds at 40 kt show over Vermont in the example above, it can be concluded that this particular wind sample exists at the 700 mb level.

A final valuable field is relative humidity, which indicates how close the air mass is to producing clouds (RH of about 70% or greater) and precipitation (RH of 90% or more). Other measures of moisture such as dewpoint or mixing ratio are not as useful since they give no indication how saturated the air is.

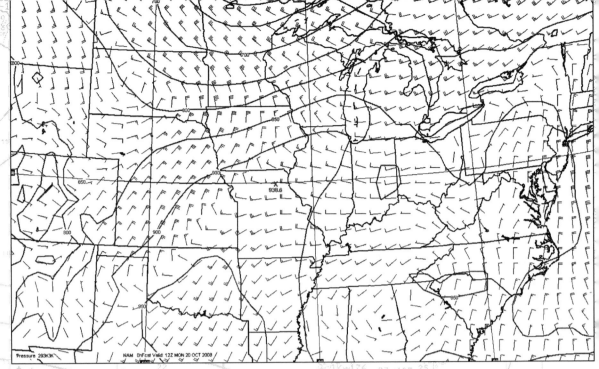

Above: Isentropic analysis for the 293 K isentrope associated with a cold front moving through the Midwest states. At this level, air was flowing from a height of about 700 mb over Minnesota southward to 900 mb in Iowa, suggesting a descent from about 10,000 ft to 3,000 ft.

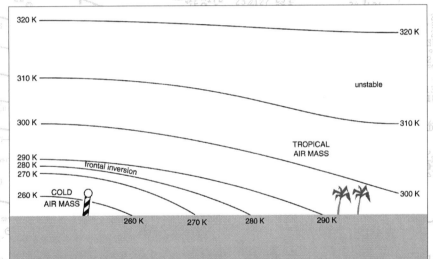

Center: Cross section of isentropic surfaces across cold air mass (left) and warm air mass (right). Surfaces are higher over cold air and lower over warm air. Whether lift or subsidence is taking place depends on the wind flow with respect to these surfaces.

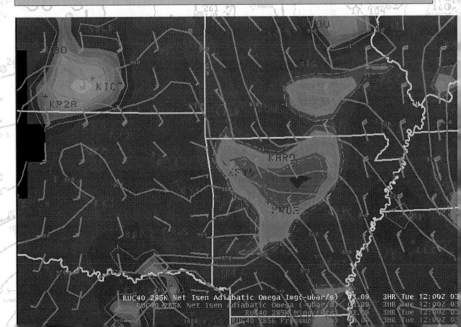

Right: Isentropic analysis as displayed on the AWIPS workstation to NWS forecasters.

Vorticity analysis

Vorticity is a measure of the spin of a given parcel of air. This is created by two qualities within the flow: shear and curvature. Shear is the difference in winds on opposite sides of the parcel. Curvature describes any turning path that the parcel must follow.

Consider a truck driver towing a giant carousel on a flatbed trailer. The carousel is free on its ball bearings, and any object or gust of wind causes it to spin. If the truck overtakes a slow driver in the left lane, while a fast car rushes by in the right lane, the carousel will tend to spin counterclockwise. This represents shear. If the road bends to the left, the carousel will spin a little faster counterclockwise while it goes around the bend. The effect would be readily observable to a person watching from an airplane. This type of turning motion represents curvature.

For most conventional vorticity analysis, forecasters use a quantity called absolute vorticity. This is the actual spin of the parcel of air, plus the spin of the earth. Another type of vorticity called relative vorticity exists, which consists of pure spin and curvature, however in practice it is only used for charts close to the surface to help find small-scale cyclones.

Absolute vorticity is typically calculated by computer systems at the 500 mb (18,000 ft MSL) level. Vorticity advection occurs where wind speed is strong (height contours close together) and vorticity gradient is strong (vorticity contours close together). Therefore where the smallest "boxes" are formed by the vorticity and height contours, advection is strongest. It uses a small part of the omega equation for vertical motion to make limited assumptions about ascent or descent.

Cyclonic vorticity advection (CVA) is tied to ascent of air, which results in clouds and precipitation. On the other hand, anticyclonic vorticity advection (AVA) is associated with subsidence, which favors clear skies. In the Northern Hemisphere, CVA is the same thing as positive vorticity advection (PVA) since higher values of vorticity are being advected in, while AVA is referred to as negative vorticity advection (NVA).

The principle of tying vorticity advection to vertical motion goes on several shaky assumptions. The first is that the atmosphere is responsive to restoration of geostrophic balance. The second is that the vorticity advection increases with height, which is often the case but not always. The third is that CVA is not negated by cold advection, or that AVA is not negated by warm advection, which is actually a common occurrence.

a closer look

The use of vorticity analysis for finding vertical motion involves several assumptions and should be used with extreme caution.

Absolute vorticity includes the rotation of the earth, whose contribution increases with latitude. It is used in the vast majority of synoptic-scale vertical motion diagnosis.

Relative vorticity is pure wind field rotation and its greatest use is simply for finding circulations on mesoscale charts, especially at low levels and at the surface.

In the northern hemisphere, PVA (positive vorticity advection) is the same thing as CVA (cyclonic vorticity advection. Likewise, NVA (negative vorticity advection) is the same thing as AVA (anticyclonic vorticity advection).

Areas of CVA are often associated with ascent. The ascent will be enhanced if warm advection is occurring in the lower troposphere. It will be negated (or even result in subsidence) if cold advection is occurring.

Areas of AVA are associated with subsidence. The descent will be enhanced if cold advection is occurring in the lower troposphere. It will be negated (or even result in ascent) if warm advection is occurring.

Areas of strong CVA should be shaded red. Areas of strong AVA should be shaded blue.

If you do not feel adept at finding areas of CVA and AVA, look for the boxes formed by the cross-overlap of height and vorticity.

websites

weather.noaa.gov/fax/nwsfax.html
weather.cod.edu/forecast
weather.unisys.com/eta
www.rap.ucar.edu/weather/model
www.wxcaster.com/models_main.htm

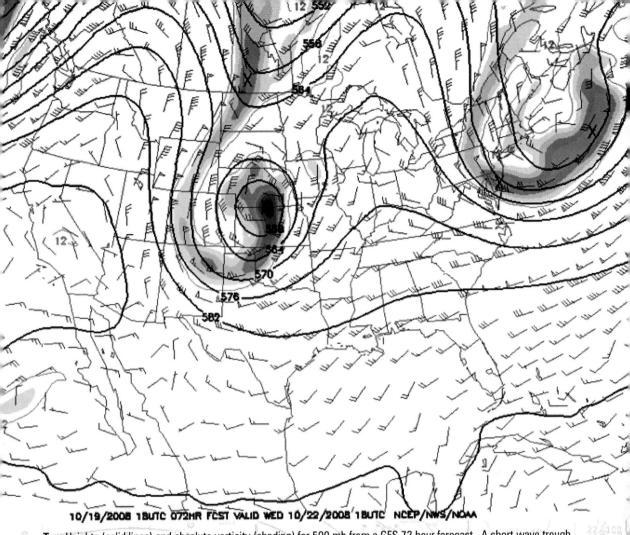

10/19/2008 18UTC 072HR FCST VALID WED 10/22/2008 18UTC   NCEP/NWS/NOAA

Top: Heights (solid lines) and absolute vorticity (shading) for 500 mb from a GFS 72 hour forecast. A short wave trough, manifested by high vorticity, is present in Kansas, with the highest value along the Kansas-Nebraska border where speed shear couples with rotation at the upper level low center. A shear lobe trails northward into Wyoming and North Dakota, with very little advection across the lobe (thus, little CVA or AVA and little vertical motion). Vorticity advection is somewhat stronger in Manitoba. *(NCEP)*

Right: Analysis of basic features in a 500 mb height-vort chart on the morning of September 11, 2001. Short wave troughs are drawn with solid line, with short wave ridges in a zigzag line. A strong area of positive vorticity advection is shaded on the right side of the chart in the Canadian maritimes region. *(NOAA)*

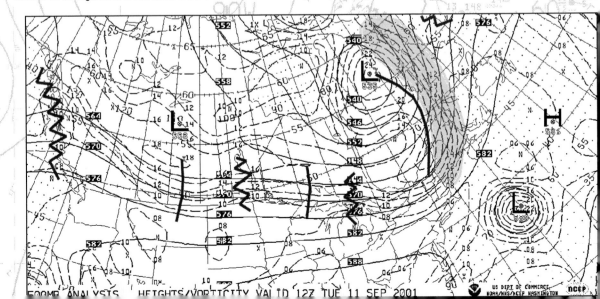

500MB ANALYSIS   HEIGHTS/VORTICITY VALID 12Z TUE 11 SEP 2001

Omega diagnostics

One meteorological parameter poses a dilemma: it is the most important quantity in operational forecasting, yet it cannot be directly measured. It is vertical velocity. A proxy exists for estimating vertical motion, called the quasi-geostrophic omega equation, which seeks to estimate the vertical velocity indirectly using other types of measurements.

The reason that vertical velocity cannot be measured directly is partly because anemometers cannot simply be mounted sideways. The vertical motion which produces widespread areas of light rain and snow, for example, may be on the order of millimeters or centimeters per second! Any attempt to measure the quantity would be masked by the background wind and turbulence. But over many hours, this motion adds up to kilometers of ascent.

Omega specifies that when cyclonic vorticity increases with height and/or warm advection is present, upward motion should occur to restore the atmosphere to geostrophic and hydrostatic balance. The amount of each quantity is proportional to the vertical motion.

Omega is expressed in microbars per second (μb/s) of vertical motion, exactly equivalent to decipascals per second (dPa/s). A microbar is of course one thousandth of a millibar. One microbar per second in the low levels is roughly one centimeter per second of ascent or subsidence (0.022 mph). A *positive value of μb/s indicates subsidence*, while a *negative value indicates ascent.*

At first glance omega appears to deliver the prized vertical motion values that all forecasters seek. However a closer look reveals serious caveats. First, accurately measuring and gauging the data and applying them to a meteorologically representative field is not as easy as it seems. Consider that a numerical analysis gridpoint can represent the observation at either Albany, Buffalo, or a weighted combination of the two. Different objective analysis techniques can produce slightly different values at the gridpoint. Any discrepancy from an "ideal" value for the gridpoint, given the starting conditions, can amplify into large errors when the model goes forth with calculations to determine omega. Even initial numerical weather prediction panels often start with significant noise.

Furthermore the vorticity parameter and thermal advection parameter often negate each other, which reduces the magnitude of vertical motion to such a small scale that the spectre of uncertainty overshadows the results. This has serious implications considering the difficulty in obtaining a properly balanced initialization.

a closer look

Omega is equal to dp/dt (the change in pressure of a parcel over a given time), and is expressed in microbars per second.

By its definition, negative values indicate ascent and positive values indicate subsidence.

Some websites and graphics packages may reverse the sign of depicted vertical motion in an attempt to express vertical motion as change in height with time rather than pressure with time. Always inspect the fields carefully before you use a new site and determine whether a positive value means downward motion (increasing change in pressure, dp/dt, with time) or upward motion (increasing height, dz/dt, with time).

One microbar per second is approximately equal to one centimeter per second (1.97 fpm) in the lower troposphere. This is not a constant speed and is actually higher aloft.

Some typical values of lift in microbars per second. Convective speeds are provided only for comparision; model omega only accounts for large-scale ascent and does not factor in convection.

Speed	Weather
0.5	Stratus
5	Light rain
50	Heavy rain, cumulus
500	Thunderstorm
5000	Tornadic thunderstorm

Moisture is needed before ascent can produce either clouds or precipitation. The higher the relative humidity, the more probable this will occur, so it is common practice to blend this output with the relative humidity at that layer.

websites

weather.cod.edu/forecast
weather.unisys.com/ngm

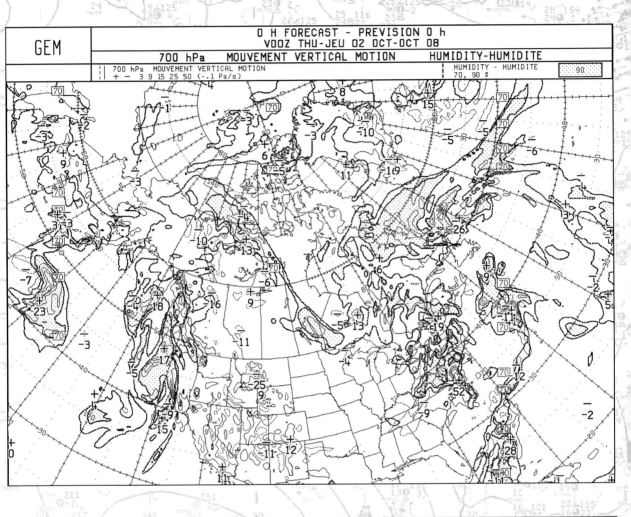

700 hPa MOUVEMENT VERTICAL MOTION
+ − 3 9 15 25 50 (−.1 Pa/s)

Top: 700 mb vertical velocity (omega) product from the Meteorological Service of Canada, as output from the initial panel of their GEM/Regional run. *(CMC)*

Right: 700 mb vertical velocity (omega) product from the RUC run, as obtained at College of DuPage's Nexlab. Omega fields are often noisy, especially in convective weather regimes and during the evening hours. *(College of DuPage)*

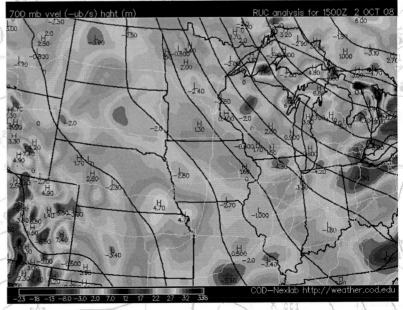

Q vector diagnostics

Q vector is a diagnostic product, with Q representing the term "quasi-geostrophic". A Q vector is an imaginary value that is equal to the rate of change of the horizontal potential temperature gradient. It gives an idea of the type of flow that must develop for the thermal wind balance to be maintained. This may consist of convergence, divergence, and most importantly, vertical motion.

The Q vector technique is a rather new addition to the meteorological toolbox, dating to 1978 in a paper published by British meteorologists Brian Hoskins, Ion Draghici, and Huw Davies. It was later advanced in the United States during the 1980s by analysis expert Stanley Barnes, and by the 1990s had gained attention as a promising tool to diagnose vertical motion. Unfortunately the attention given to this technique has been lukewarm, but there is probably no other technique that's as useful for estimating large-scale vertical motion at various levels without the noise inherent in omega fields.

Looking at one Q vector by itself, whenever its magnitude is high, a strong horizontal ageostrophic wind is implied. This is interpreted as a response by the atmosphere to restore the thermal wind balance. Vertical motion is likely to develop to compensate for the imbalance when the Q vectors are convergent or divergent.

In a convergent pattern, the Q vectors tend to point at one another. This implies ascent at that level. In a divergent pattern, the Q vectors point away from one another. This implies subsidence at that level. Some panels and software displays are able to directly measure the magnitude of convergence or divergence, eliminating much of the guesswork. The result immediately suggests the sign and the intensity of the implied ascent or descent.

Also, Q vectors are often overlaid on top of a product showing isotherms for that level, or thickness for a layer centered on that level. This reveals important information about whether the thermal boundaries and fronts are strengthening or weakening. When Q vectors point from cold to warm air, the thermal gradient is strengthening and frontogenesis is implied. When Q vectors point from warm to cold air, the thermal gradient is weakening and frontolysis (weakening of a front) is implied.

Going a step further, the thickness gradient pattern has been shown to have some bearing on the surface pressure patterns. Where the thermal gradient is showing an "S" shape and Q vectors indicate frontogenesis, cyclogenesis is implied, with falling surface pressures and deteriorating weather.

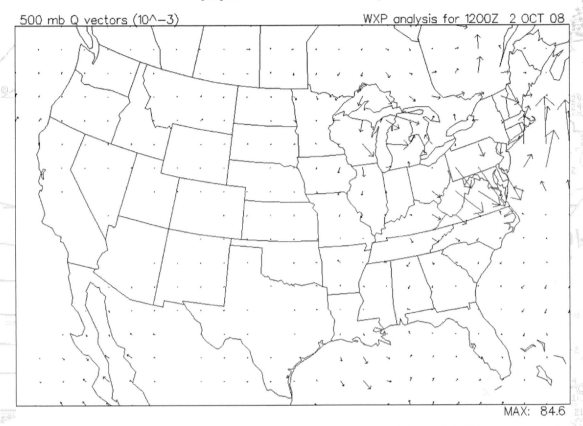

500 mb Q vectors (10^-3)   WXP analysis for 1200Z 2 OCT 08

MAX: 84.6

Top: Q vector analysis as obtained from Plymouth State. This layer shows a single-level Q vector field at 500 mb. The TTU charts stand out among a vacuum of Q-vector products on the Internet. No known sources of layer Q vector charts were known at press time, though such fields can be generated with standalone software. *(Plymouth State)*

Right: A companion field available on the Plymouth State site actually measures the Q-vector divergence. The fields are extremely noisy, illustrating one problem with literal interpretation of the data without looking at the actual Q vectors. *(Plymouth State)*

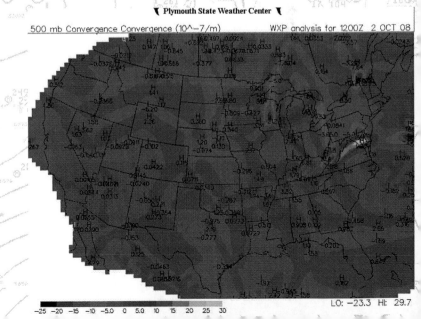

500 mb Convergence Convergence (10^-7/m)   WXP analysis for 1200Z 2 OCT 08

LO: -23.3 HI: 29.7

-25 -20 -15 -10 -5.0   0   5.0   10   15   20   25   30

Thermodynamic diagram

The thermodynamic diagram shows the profile of temperature and dewpoint above a given weather station with height. The X-coordinate is always temperature, and the Y-coordinate is height. Using these coordinates, observations of temperature and dewpoint at various heights are plotted. Since dewpoint is always equal to or lower than temperature, the dewpoint trace is usually to the left of the temperature trace. It is typically shown as a dashed rather than a solid line.

In the *skew T diagram*, widely used by North American meteorologists, the temperature coordinates are skewed 45 degrees to the right to help make changes in the slope of the temperature profile easier to see. In other words, the background coordinates of height and temperature do not form a grid like classic graph paper; rather the vertical lines are turned diagonally by about 45 degrees to the right.

The most important use of the sounding is to assess the degree of instability present in the atmosphere. If any part of the sounding leans sharply to the left with height, it is assumed to have a large lapse rate (a great temperature decrease with height). If the dewpoint trace indicates significant moisture in the lower levels, the moisture and its potential for latent heat release will combine with this large lapse rate to produce an unstable atmosphere.

The simplest measure of instability is the Showalter Stability Index (SSI) and the Lifted Index (LI). However, both of these rely on a simple comparision of the parcel with the environment at one level. A far more accurate measure is Convective Availability of Potential Energy (CAPE), which assesses the parcel-environment temperature differential throughout the entire vertical column. A CAPE value should always be used except when it is not available. All instability calculations depend on an accurate representation of parcel temperature and moisture, which in turn requires a representative integration of low-level moisture and an accurate temperature forecast. Human manipulation of the parcel attributes are always worthwhile, and are easy to do on paper SKEW-T diagrams and in certain weather software applications.

The basic concept for determining the type of precipitation that reaches the surface is to begin in the mid-levels of the troposphere, determine what type of precipitation it begins as, and observe the temperature regimes that affect the particle as it falls downward. As a rule of thumb, 1200 ft of warm air is considered ample to completely melt snow, with 400 ft of cold air considered enough to freeze liquid precipitation.

a closer look

The thermodynamic diagram is used for establishing the character of the entire air mass above a given station. It consists of a plot of temperature, moisture, and wind.

Height lines are horizontal and are usually calibrated in millibars. The top of the chart is usually 100 mb (about 53,000 ft) and the bottom is usually 1050 mb (about minus 300 ft MSL).

Temperature lines slope up and to the right. They are calibrated in degrees Celsius with an interval of every 10 C°.

Dry adiabats slope up and to the left. These indicate how a dry parcel will cool as it rises.

Moist (wet) adiabats slope upward and then curve sharply to the left. These indicate how a saturated parcel will cool as it rises.

Mixing ratio lines slope upward and to the right. They are more vertical than the temperature lines, and are typically omitted in the upper portion of the sounding. A rising parcel's dewpoint will follow these lines upward until it saturates.

A software package dedicated to working with soundings is highly recommended and makes quick, accurate work of parcel estimates and daytime heating influences.

The "bible" for the SKEW-T is "The Use of the Skew-T Log-P Diagram in Analysis and Forecasting" by Robert C. Miller, AWS TR-200, published in 1972 and changed very little since then.

websites

www.rap.ucar.edu/weather/upper
weather.unisys.com/upper_air/skew
http://weather.cod.edu/analysis
weather.uwyo.edu/upperair/sounding.html

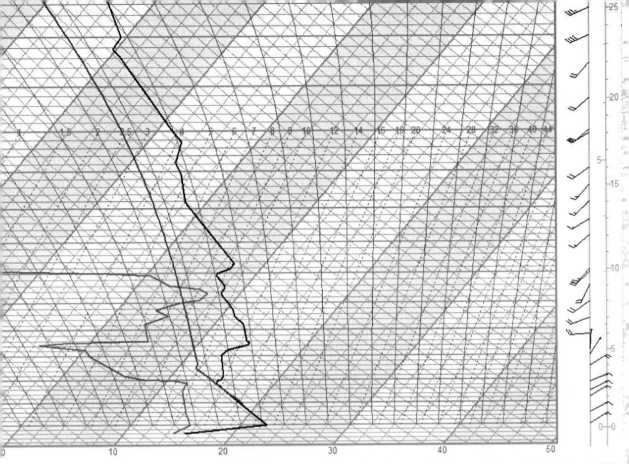

Above: Detailed sounding generated by the Windows analysis software Digital Atmosphere. Detailed diagrams like this are useful in meticulous severe weather forecasting work. Note the hypothetical parcel lift outlined in red, which is a parcel constructed from the mean mixing ratio and mean potential temperature in the lowest 150 mb of the atmosphere.

Right: Sounding from the UCAR weather page. It shows a frontal inversion near the surface, about 1000 ft deep, and an elevated layer of moisture, probably resulting in mid-level clouds, from the 700 to 400 mb layer. Instability indices are placed at the top of the diagram. At the top left is a hodograph.

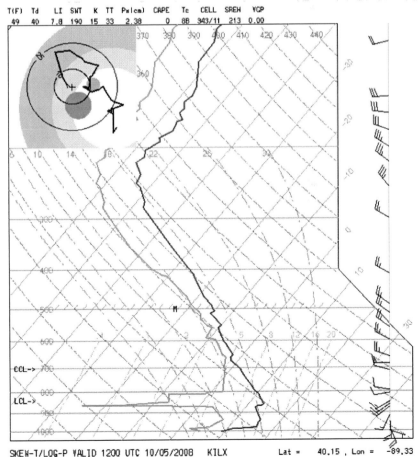

T(F)	Td	LI	SWT	K	TT	Pw(cm)	CAPE	Tc	CELL	SREH	VGP
49	40	7.8	190	15	33	2.38	0	8B	343/11	213	0.00

SKEW-T/LOG-P VALID 1200 UTC 10/05/2008   KILX   Lat =   40.15 , Lon = -89.33

Wind profiler

A wind profiler is a special type of Doppler radar that measures wind speed with a fixed antenna. The radar is able to project in two perpendicular locations off the vertical axis in order to measure the horizontal wind components. The advantage of a wind profiler is that it can sample winds continuously without the quantity of scatterers needed by a WSR-88D producing a VAD/VWP product. They have fewer moving parts, making them more reliable.

Profilers operate in a high mode and a low mode. The low mode usually covers the lower and middle troposphere (up to about 9 km), while the high mode covers the upper troposphere (from about 7 to 16 km) in a much more sensitive detection mode.

A piggyback technology incorporated in most soundings is called *Radio Acoustic Sounding System (RASS)*. Its main function is to estimate temperature at various layers. Since the speed of sound is dependent on temperature, RASS allows temperature to be measured. A 900 Hz burst, one octave above the musical middle A, is transmitted upward into the atmosphere and its echo is measured. This data is not as accurate as radiosonde measurements so it should be used with caution in convective forecasting.

A special type of wind profiler is called *Sonic Detection and Ranging (SODAR)*. It relies entirely on acoustic echoes for detecting wind speed, and its vertical limit of about 1000 ft confines it to research and other special applications.

Europe has continued to invest in wind profiler technology and currently maintains a network of about 26 wind profilers stretching from northern Italy and Spain to Germany, the United Kingdom, and Norway. The data is available on the Internet without restriction at the UK Met Office website <*www.metoffice.gov.uk/research/interproj/cwinde/profiler*>.

Profiler data is usually viewed with time-height profiles. When using a new website, meteorologists should always study the coordinate system since some websites put the newest data on the left side rather than the right one. Color coding of wind plots may also indicate either wind speed or RMS error; there is no standard. It is also possible to display data as a geographical plot across a region, yielding excellent information on changes in the wind field over a period of hours.

Wind profiler data can be supplemented with WSR-88D VAD/VWP wind profiles, which are not as accurate but very similar in forecast use.

a closer look

Wind profilers are a special type of clear-air Doppler radar with no moving parts. They often come with RASS units which measure temperature by projecting acoustic beeps toward the zenith.

Wind profilers usually require six minutes for each cycle. This is comprised of three 2-minute samples: one at the zenith, one tilted in the X direction, and one tilted in the Y direction. Within each of these modes, there is a 1-minute sample in low mode and another in high mode (see text).

The NOAA 404 MHz wind profilers are programmed to shut down when SARSAT search and rescue satellites pass overhead, in order to avoid interfering with their sensors which operate on a similar frequency. This inhibit mode occurs about 4 to 10 times per day, and lasts for about six minutes. It will cause a gap in the observations. The NOAA FSL Profiler site maintains a list of scheduled inhibit times.

Strips of bad data may be caused by aircraft flying over the profiler.

One overlooked method of viewing profiler data is using horizontal plots of multiple stations at a given level. This can fill in data between radiosonde stations.

Wind profilers in the United States have often ended up on the chopping block during announced budget cuts, only to find new funding. In 2004 it was expected that the national network would go entirely offline. Fortunately this scenario did not pan out.

websites

www.profiler.noaa.gov/jsp/profiler.jsp
www.rap.ucar.edu/weather/upper/
weather.cod.edu/analysis
www.met-office.gov.uk/research/interproj/
  cwinde/profiler/

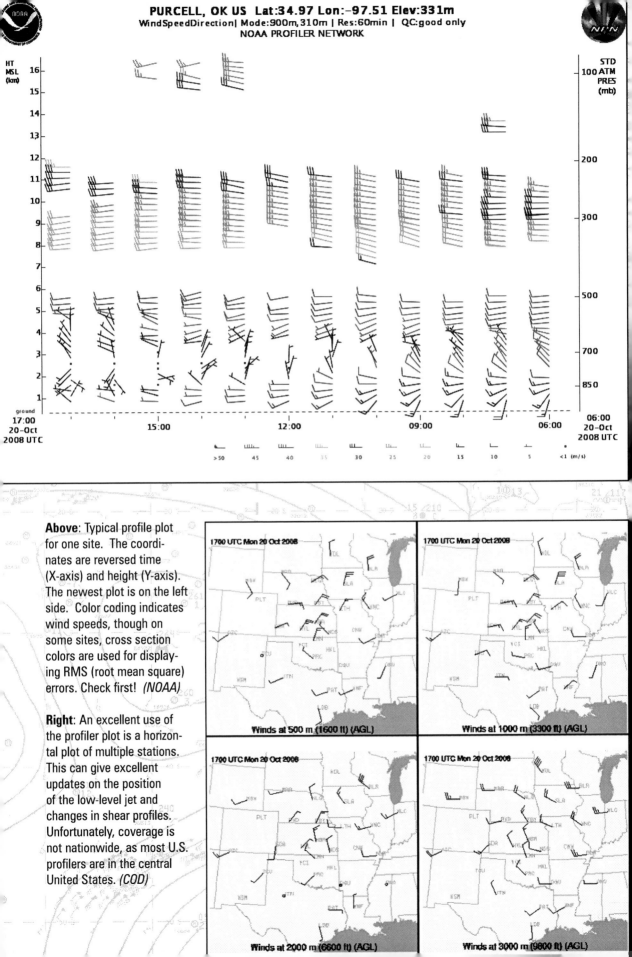

PURCELL, OK US Lat:34.97 Lon:-97.51 Elev:331m
WindSpeedDirection| Mode:900m,310m | Res:60min | QC:good only
NOAA PROFILER NETWORK

Above: Typical profile plot for one site. The coordinates are reversed time (X-axis) and height (Y-axis). The newest plot is on the left side. Color coding indicates wind speeds, though on some sites, cross section colors are used for displaying RMS (root mean square) errors. Check first! *(NOAA)*

Right: An excellent use of the profiler plot is a horizontal plot of multiple stations. This can give excellent updates on the position of the low-level jet and changes in shear profiles. Unfortunately, coverage is not nationwide, as most U.S. profilers are in the central United States. *(COD)*

Winds at 500 m (1600 ft) (AGL)

Winds at 1000 m (3300 ft) (AGL)

Winds at 2000 m (6600 ft) (AGL)

Winds at 3000 m (9800 ft) (AGL)

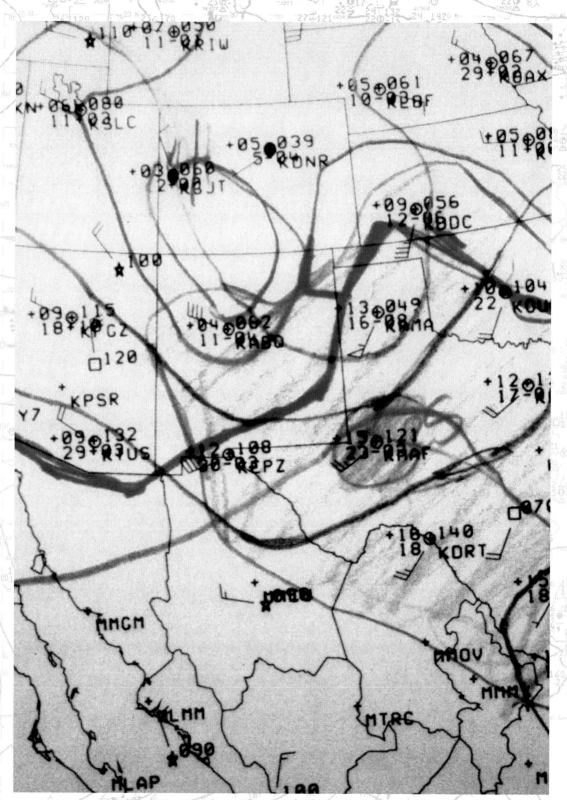

Hand analysis of 700 mb conditions photographed on the operations desk at WSFO Fort Worth during a severe weather outbreak in 2003. *(Tim Vasquez)*

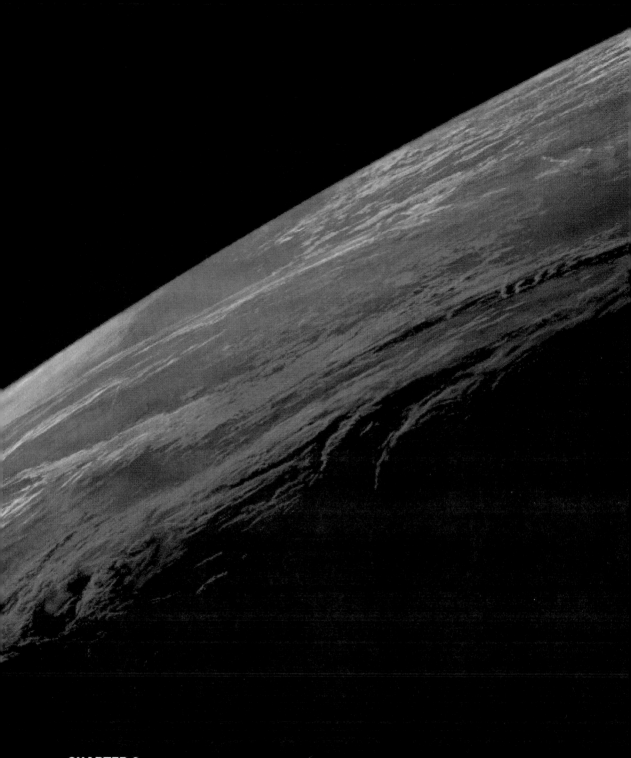

CHAPTER 2

SATELLITE

Satellite

In April 1960, the world's first weather satellite was launched, named TIROS (Television InfraRed Observation Satellite). This system was little more than a television camera mounted on a small orbiting platform. It was only during the 1970s when satellite technology began reaching today's levels of sophistication, yielding sharp hourly images. And finally in the mid-1990s, detailed satellite imagery became available freely to anyone who had access to the Internet.

Weather satellite imagery consists primarily of visible, infrared, and water vapor imagery. There are also complex multispectral products available. Each will be covered in the sections ahead. First, we will take a look at the major classes of weather satellites: *geostationary* and *low-orbiting*.

Geostationary earth orbiting (GEO)

Geostationary satellites are placed in orbit 19,323 nm (22,236 sm or 35,786 km) above sea level. At this altitude it is possible to put an object in a semipermanent "free fall" orbit around the earth with an orbital speed that coincidentally matches the earth's rotation. By doing this, the weather satellite can be parked above a given spot on earth.

The weather satellite imager scans the Earth in a raster format, using the spacecraft's spin for imaging along one axis and a stepping mirror for incrementing the imager along the perpendicular axis. This allows the spacecraft to build an image with a minimum of moving parts.

The first geostationary satellite was ATS-1, launched in December 1966. The first of the GOES-class weather satellites was SMS-1, launched on May 5, 1974. It was operational from 1974 to 1981. SMS-1 was initially a NASA project but was turned over to NOAA oversight. Since then, 13 dozen geostationary satellites have been launched by the United States, and are referred to as GOES (Geostationary Operational Environment Satellites).

Japan and Europe both launched their first geostationary weather satellites in 1977, maintaining consistent reliable coverage ever since. The Japanese satellite is known as GMS, while the European satellite is called METEOSAT. India, Russia, and China have launched geostationary weather satellites for their parts of the world with varying degrees of success and failure.

Low Earth orbiting (LEO)

The term "low-Earth orbiting" is a name

Preceding page: Artistic image of GOES-10 view, showing an August evening falling on British Columbia and Yukon. *(NASA/GSFC)*

Below: One of the first TIROS images: May 1960, showing a line of severe thunderstorms moving through northern Missouri. Satellite imagery would revolutionize meteorology during the coming decades. *(NOAA)*

given to weather satellites which orbit only about 500 miles above the ground. They are often called *polar orbiters*. A rotation is completed about once every 1.5 hours, and the orbit is typically *sun-synchronous*, which means that if the Earth's rotation was completely stopped with respect to the sun and the satellite's path was marked on a world map, the spacecraft would never deviate from that path. The Earth's rotation underneath the sun-synchronous orbit allows different areas of the Earth to sweep by underneath, each part passing underneath every 12 hours: once on the day side and once on the night side. Polar orbiters have an inclination that takes them over polar regions, giving them unprecedented ability to image polar regions such as Alaska, northern Canada, and Antarctica, all of which are seen at too much of a slant on GOES satellites to give useful images.

Nearly two dozen civilian LEO satellites have been launched by the U.S. government. They are all referred to as the POES (Polar Orbiter Environmental Satellite) series, and are comprised of satellites designated "NOAA" (older ones were known as TIROS). The U.S. military also operates a network of LEO weather satellites known as DMSP (Defense Meteorological Satellite Program), however the imagery is encrypted and is shared with the public only through special research programs. The DMSP series carries night vision sensors that allow visible weather imagery to be obtained in moonlit conditions.

NASA launched the MODIS (Moderate-Resolution Imaging Spectroradiometer) system on board the Terra and Aqua satellites in 1999 and 2002, respectively. Terra and Aqua are LEO birds. The result is LEO weather imagery with extremely high resolution (0.25 km). The images are distributed solely at <*rapidfire.sci.gsfc.nasa. gov/realtime*>. A significant disadvantage of this system is that can take 1 to 10 hours to receive usable images, depending on the system load.

Though LEO imagery has the advantage of mapping anywhere on earth without slant angle degradation in the polar regions, it does have some drawbacks. It can only image the same spot on earth twice a day, and the view always requires complicated georeferencing (adding borders and geography). The images also tend to have some distortion since the satellite is so low. It should be noted that the NOAA LEO series has much

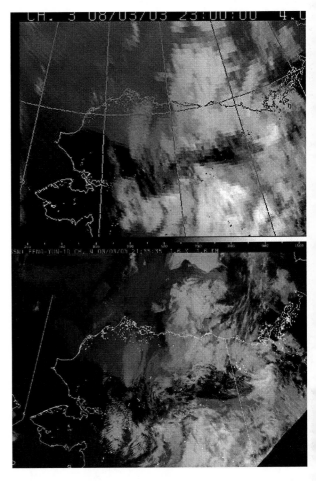

Above: Highest possible infrared resolution for GOES-10 geostationary satellite (top) compared to equivalent Chinese Feng-Yun 10 polar orbiter (bottom). The degradation in the GOES imagery caused by slant, combined with the 4 km infrared resolution in GOES compared to the 1 to 2 km packages in the polar orbiters, makes polar orbiters a staple for polar forecasting. *(NOAA/ARH)*

better infrared resolution than that onboard the GOES satellite.

Imagery on the Web

Most real-time satellite photos available on the Internet are of very high quality resolution, and available almost anywhere in the world. From the GOES satellites, North America receives 1-km resolution visible imagery every 5 to 15 minutes; 4 km infrared imagery every 15 minutes; and 8 km water vapor imagery every 15 minutes. This is helped only by 1 km infrared imagery by low-Earth orbit satellites.

Elsewhere, hobbyists and some weather consulting firms in Europe and Asia are heavily stifled by 4 km or 8 km resolution barriers imposed by the owners of the METEOSAT and GMS satellites. This leaves a very limited assortment of satellite data to choose from, with poor suitability for mesoscale purposes. Polar orbiter data from the United States, China, and Russia provides important supplementary images.

An excellent directory of real-time data is at <www.nwas.org/committees/rs/nwasat.html>. For a selection of images for Alaska that includes POES and FY imagery, see <www.arh.noaa.gov>. Official websites for government weather satellite programs include <www.goes.noaa.gov>, <www.oso.noaa.gov/poes>, <smis.iki.rssi.ru>, <www.eumetsat.de>, <mscweb.kishou.go.jp> and <www.imd.ernet.in>.

Right: First full-disk image from the GOES series of satellites. This image was taken in October 1975 from GOES-1. NASA had operated the ATS and SMS demonstrator series over a period of nearly ten years before this date, but the GOES satellite series marks the first of a series of satellites designed especially for NOAA. *(NOAA)*

Below: The world's first geostationary satellite image. Dating back to 11 December 1966, this image was taken by ATS-1, a satellite operated by NASA. This satellite was used to relay color television and White House communications across the continent. Early weather satellite technology was largely hindered by a lack of suitable display technology at each field office. Television screens were far too coarse, high-definition computer monitors were still 15 years away, and laser printers would not be invented until 1975. Therefore film recorders were a mainstay in the early years. Quality and economy were not impressive. *(NOAA)*

GOES-1 DPT 298 1645Z 25 OCT. 75

Visible imagery

Visible satellite imagery is the most intuitive type of satellite imagery, as it detects exactly what an astronaut would see from space. Visible imagery measures brightness, which is a function of illumination and albedo. Albedo is the percentage of incoming sunlight which is reflected into space.

Clouds composed of water droplets have a much higher albedo than ice crystal clouds. By contrast, infrared imagery is sensitive to temperature and readily detects ice crystal clouds. So cirrus and cirrostratus layers may appear thinner on visible imagery than infrared imagery suggests. On the same token, fog and stratus usually has the same temperature as the ground, making it hard to find on infrared imagery, but at dawn when visible imagery is available the high albedo of fog and stratus makes it immediately identifiable.

There are several important concepts to remember when analyzing visible satellite imagery.

Visible imagery is not available at night. The military DMSP satellite offers night-vision visible imagery if moonlight is available, however these are generally not distributed.

Cloud type can be easily assessed, which yields considerable qualitative information on the state of air masses in a forecast region.

Visible imagery is rarely used on television weathercasts, which in recent years have begun leaning heavily toward processed products. Infrared imagery offers 24-hour consistency.

Wind direction in the low levels can be determined simply by looking at the alignment of cumulus fields. This is difficult to detect with lower-resolution infrared imagery.

Dust plumes can be detected. They take on a distinct smudgelike appearance, and are not well resolved by infrared imagery.

Benign or subtle low-level boundaries, such as cold fronts and thunderstorm outflow boundaries, are readily seen on visible imagery. Such boundaries are rarely detectable on infrared imagery due to their coarser resolution; at night, radar is preferred over infrared imagery.

Right: Terra MODIS imagery at 1710 UTC, 12 September 2008, showing the Texas coastline as Hurricane Ike approaches. Most prominent are the bands of towering cumulus and stratocumulus, oriented into cyclonic streets. Outflow cirrus covers much of the lower right half of the image. *(NASA Global Hydrology and Climate Center)*

Infrared imagery

Infrared imagery looks a lot like visible imagery, but it is actually temperature we are seeing. Also there's a significant difference: infrared imagery is available 24 hours a day, whereas visible imagery goes away at night.

Dark areas correspond to areas of high thermal radiation, while white areas indicate areas of low thermal radiation. Therefore anything dark is warm, and anything white is cold. In fact, since infrared imagery helps to measure emitted radiation, it is possible to use a scale to find the temperature of any pixel in an image! For decades, weather offices in citrus farming districts have even looked for specific color ranges matching freezing temperatures.

Infrared imagery is often enhanced, which refers to the technique of adding false color banding or artificial color gradients to allow certain temperature ranges and cloud patterns to stand out. The most basic enhancement was known by decades as the ZA curve, which simply increases contrast . A more sophisticated enhancement is the MB curve, used by the National Weather Service since the 1970s to highlight temperatures below minus 32°C in various shades of gray and extract detail from cold, amorphous stratiform tops. Various Internet sites often implement their own enhancement schemes. Some enhancement schemes, including the example at the right, use a derivation of the MB curve.

Infrared satellite imagery from the GOES satellites is no better than 4 km in resolution. As most visible satellite imagery uses 1 km resolution, this means that there is usually only one infrared pixel for every 16 visible pixels. Therefore in many cases it is better to use visible imagery when available. The exception is polar orbiter imagery, which can image infrared at 1 km resolution. These can be used for case studies and special weather events when the best possible infrared imagery is needed.

Clouds may be invisible on infrared imagery when they take on the same temperature as the ground. This is true of fog; the full extent of fog is often not known until the morning hours when visible imagery is available. Stratus can often be invisible when the ground is cooler than normal. A technique that forecasters often use is to monitor whether lake surfaces, which are warmer, remain visible. A warm lake that "disappears" indicates that it has been obscured by a cloud layer.

Circular, bubble-like forms with extremely cold signatures are usually convective showers and thunderstorms. They are some of the most prominent objects, even on the crudest infrared photos.

a closer look

Infrared imagery is valuable for monitoring the hourly progress of weather systems round the clock and for assessing the intensification or weakening of convective systems.

Infrared imagery is unable to monitor low stratus and fog effectively, nor can it easily find cloud layers in arctic air masses.

GOES infrared imagery has a resolution of 4 km and operates on a wavelength of 11 microns.

Polar orbiter AVHRR infrared imagery has a resolution of 1 km (much better than GOES) and is provided on data channel 3 (shortwave) and 4 (longwave). Channel 2 is a near-infrared channel.

During arctic air mass outbreaks, the temperatures in the northern United States and Canada can be so cold that at first glance, the infrared imagery seems to show a broad sheet of cirrostratus. You may mistake cirrostratus for the ground, or may be unable to identify any clouds whatsoever.

Near infrared imagery (near IR) is a special type of infrared imagery that uses the 3.9 micron band. It shares a lot in common with visible imagery and is used to differentiate low-level features that are often masked, such as snow, fog, and stratus.

Multispectral images combine infrared and even visible channels in different ways. These can give information about cloud heights, moisture, clouds, and fog.

websites

wwwghcc.msfc.nasa.gov/GOES
www.rap.ucar.edu/weather/satellite
www.ssec.wisc.edu/data
weather.cod.edu/analysis

-70 -68 -66 -64 -62 -60 -58 -56 -54 -52 -50 -48 -46 -44 -42 -40 -38 -36 -34 -32 -30 -28 -26 -24 -22

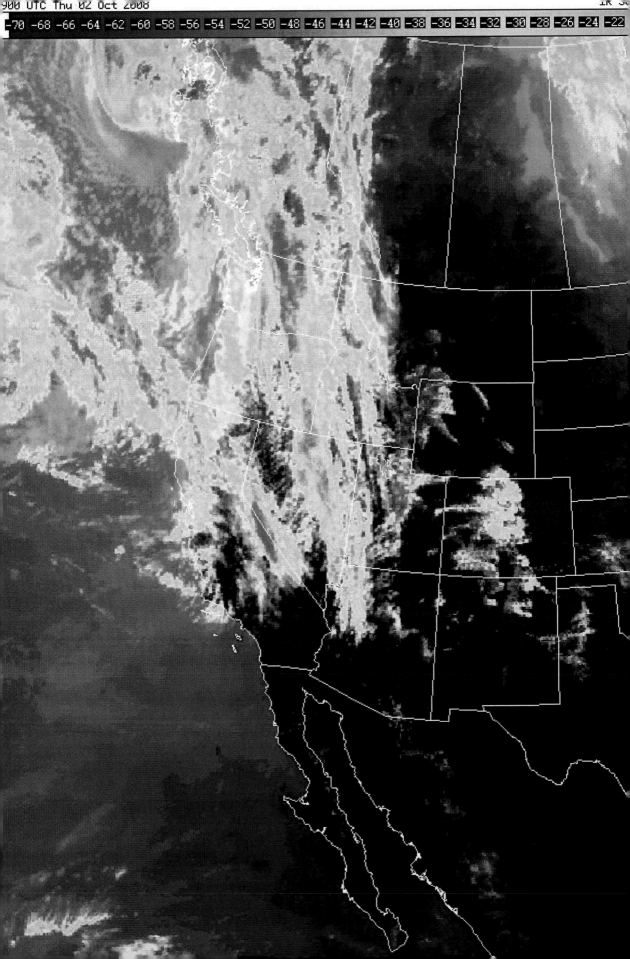

Water vapor imagery

Hang around any forecast office for long, and you'll get the idea that "water vapor imagery" and "moisture" go hand-in-hand. This is absolutely correct. The radiosonde observation network only detects small pieces of the moisture field across a given continent, and this makes water vapor imagery a powerful tool for filling in the blanks in a visual sense. The assumption is that moisture is associated with synoptic-scale ascent, while dry areas are associated with synoptic-scale subsidence.

Water vapor, particularly in the mid troposphere, tends to absorb radiation in the 6 to 7 micron band. Therefore the satellite imagery does not look directly for water vapor but rather suggests areas where radiation may have been absorbed by vapor in the atmosphere. The 6.7 micron band is picked up by special mercury cadmium telluride imagers on the weather satellite. As with infrared imagery, the scale is inverted so that strong radiation takes on a dark appearance while weak radiation looks bright.

Where a bright area is shown, it suggests the presence of moisture. This is actually a weak radiation signature, suggesting that enough moisture was present to absorb 6.7-micron radiation from the surface. Thus we can conclude that moisture exists in the mid-levels of the atmosphere. However this weak radiation signature can also come from a cold region, which does not emit much radiation at all to begin with.

A dark area indicates that all the radiation from the surface made it to the satellite without being attenuated. Thus we can conclude that the earth's surface is warm and that no moisture was present to absorb the radiation as it moved upward. Thus the atmosphere is probably dry.

With this in mind, there are some very important limitations with water vapor imagery:

It only works well in the middle troposphere, mostly between 350 and 650 mb (12,000 and 25,000 ft). Therefore, whiteness (associated with moisture) can be present even though the low levels or upper levels are extremely dry.

It works best in warm atmosphere. Whiteness, the result of a lack of radiation, will also be produced in areas where it is so cold that no radiation can be emitted into space. This makes the imagery least useful in northern latitudes during the winter, where everything appears white in the 6.7-micron band.

It is degraded by the presence of clouds. Whiteness can be a layer of water droplets or ice (clouds) that are exceptionally thin, rather than a very deep, rich layer of water vapor.

a closer look

Water vapor imagery is used for assessing mid-level and upper-level moisture in warm tropospheric environments.

GOES water vapor imagery is sensed at a wavelength of 6.7 microns and has a resolution of 8 km. Polar orbiter AVHRR systems do not produce water vapor imagery.

Clouds do not need to be present to show up on water vapor imagery. All that is needed is water vapor (a gas) in the middle troposphere to create a signature. Therefore moist bands can show up well before clouds begin forming on infrared or visible imagery.

A dark pixel indicates strong 6.7-micron radiation was received and is suggestive of warm, dry conditions.

A bright pixel indicates weak radiation was received, and in turn this implies that the atmosphere at that location is moist in its middle or upper levels and/or is cold.

Water vapor imagery is excellent for picking out the position of the subtropical jet (STJ). The STJ usually lies along the poleward periphery of a broad bulge of tropical moisture in the subtropical latitudes and is most common in the winter months.

Very dark bands on water vapor imagery in the wake of a baroclinic storm system may be in proximity to a very strong jet max. This can help refine the jet max position. The location is usually on the periphery of the dark band (poleward) and the brighter area (equatorward).

websites

wwwghcc.msfc.nasa.gov/GOES
www.rap.ucar.edu/weather/satellite
www.ssec.wisc.edu/data
weather.cod.edu/analysis

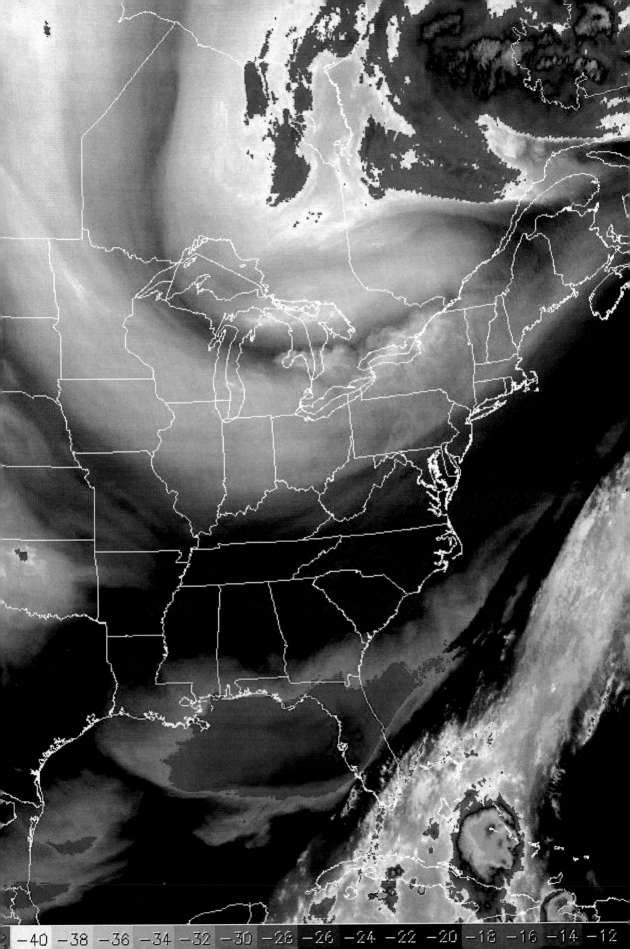

GOES sounding

Since May 1994, when GOES-8 was put into operational service, an improved radiometer package and better satellite design have made satellite-based atmospheric soundings a reality. Such soundings entered operational use in July 1995. However, the technology and science is still in its infancy. The satellite sounding products are heavily dependent on output from numerical weather prediction models. The term "GOES sounding" is somewhat of a misnomer — a better description might be "NWP/GOES sounding" to highlight that it is a fusion of the two technologies. However, this book will adhere to the standard naming conventions used in forecasting.

GOES Sounder sector scans are performed every 30 minutes, which scans the entire Earth's disk. Sensor data from 18 different infrared frequencies, ranging from 3.7 to 14.7 microns, plus the visible channel, are collected. This data has a resolution of about 8 km spaced every 10 km. This 8 km box produces an area on the Earth's surface known as a field of view (FOV), to distinguish this as a volumetric product rather than a point measurement. Each of the infrared channels are sensitive to radiation from a specific atmospheric layer. The word "layer" is an important distinction, because radiation is not sensed at individual levels but in layers. In many cases this sensitivity spans as much as 15,000 ft of depth, which has important implications for the sounding product.

For every 50 km a sounding is computed. The model sounding from the GFS run (formerly the AVN) is used as a first guess. In other words, it is assumed that the Eta sounding is what the GOES satellite is seeing. The satellite data and surface observation for the location are then automatically examined over an area as large as 50 km (ten fields of view along each axis) to see whether the column is free of clouds. Following this, radiation for the column is estimated and adjusted using surface data and GFS model variables. If the column is cloudy, a cloud top pressure is estimated, and the interrogation stops at this point. No satellite sounding can be computed.

For a clear column, the GFS sounding is adjusted so that its estimated radiation signature closely matches what the GOES satellite detects. The tweaked sounding is referred to as the "GOES sounding".

The suitability of GOES soundings to everyday forecasting problems is not yet clear. Much of the material dealing with GOES soundings has been in the arena of research or with very limited case studies, and a considerable amount of operational investigation still needs to be done.

a closer look

The GOES sounding is actually an GFS forecast sounding with adjustments made using satellite-detected radiation signatures. It is not a direct, raw measurement of a tropospheric column.

The GOES sounder is most useful in desolate areas such as oceans and gulfs.

The term "GOES retrieval" is often used to describe the process of creating a GOES sounding.

NESDIS attempted to use the Eta model for a brief period, but reverted back to the GFS after wrestling with problems with the Eta's well-documented low-level moisture biases.

The source model's temperature profile is changed little, sometimes not even perceptibly, by the GOES sounder. In many cases the resulting profile is more representative of the GFS model solution than the satellite-borne data.

The model's moisture profile may be changed significantly by the GOES sounder. These changes are very useful from a forecasting perspective, more so than thermal output from GOES soundings. In other words, in a convective situation the soundings should be used to monitor moisture rather than nit-pick cap strength.

Clouds, which radiate and absorb heat, contaminate the sample column and prevent GOES soundings from being made at a given location. Such locations are automatically skipped by sounding products.

websites

www.star.nesdis.noaa.gov/smcd/opdb/
    goes/soundings
cimss.ssec.wisc.edu/goes/realtime

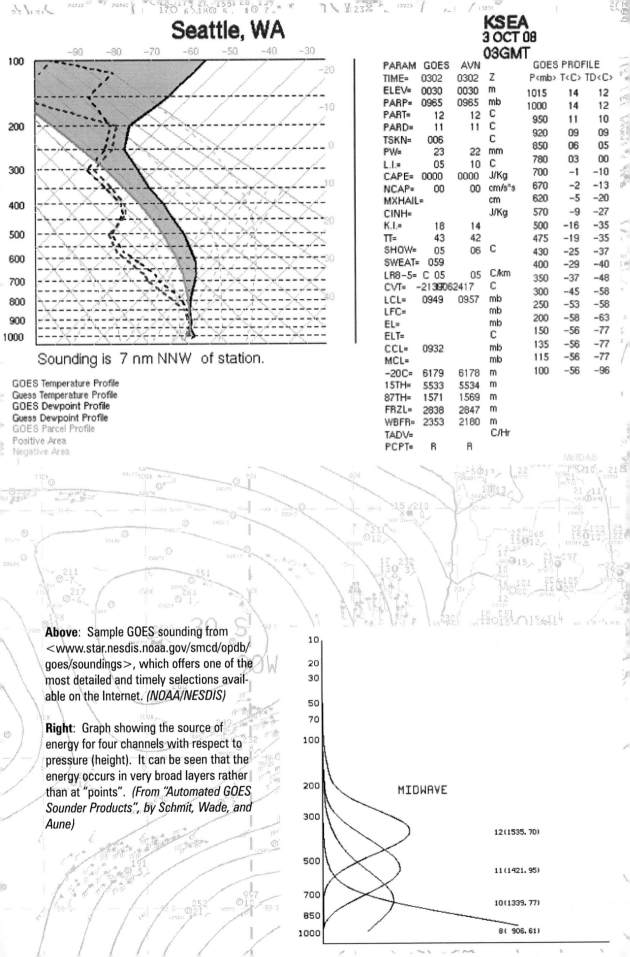

Seattle, WA

KSEA
3 OCT 08
03GMT

Sounding is 7 nm NNW of station.

GOES Temperature Profile
Guess Temperature Profile
GOES Dewpoint Profile
Guess Dewpoint Profile
GOES Parcel Profile
Positive Area
Negative Area

PARAM	GOES	AVN	
TIME=	0302	0302	Z
ELEV=	0030	0030	m
PARP=	0965	0965	mb
PART=	12	12	C
PARD=	11	11	C
TSKN=	006		C
PW=	23	22	mm
L.I.=	05	10	C
CAPE=	0000	0000	J/Kg
NCAP=	00	00	cm/s*s
MXHAIL=			cm
CINH=			J/Kg
K.I.=	18	14	
TT=	43	42	
SHOW=	05	06	C
SWEAT=	059		
LR8-5= C 05		05	C/km
CVT=	-2139062417		C
LCL=	0949	0957	mb
LFC=			mb
EL=			mb
ELT=			C
CCL=	0932		mb
MCL=			mb
-20C=	6179	6178	m
15TH=	5533	5534	m
87TH=	1571	1569	m
FRZL=	2838	2847	m
WBFR=	2353	2180	m
TADV=			C/Hr
PCPT=	R	R	

GOES PROFILE

P<mb>	T<C>	TD<C>
1015	14	12
1000	14	12
950	11	10
920	09	09
850	06	05
780	03	00
700	-1	-10
670	-2	-13
620	-5	-20
570	-9	-27
500	-16	-35
475	-19	-35
430	-25	-37
400	-29	-40
350	-37	-48
300	-45	-58
250	-53	-58
200	-58	-63
150	-56	-77
135	-56	-77
115	-56	-77
100	-56	-96

Above: Sample GOES sounding from <www.star.nesdis.noaa.gov/smcd/opdb/goes/soundings>, which offers one of the most detailed and timely selections available on the Internet. *(NOAA/NESDIS)*

Right: Graph showing the source of energy for four channels with respect to pressure (height). It can be seen that the energy occurs in very broad layers rather than at "points". *(From "Automated GOES Sounder Products", by Schmit, Wade, and Aune)*

MIDWAVE

12(1535.70)
11(1421.95)
10(1339.77)
8(906.61)

Full-disc image from the Feng Yun FY-2A geostationary satellite manufactured, launched, and operated by China. This image shows Asia on 23 December 1997 in a combination of visible and infrared. Until 1997 nearly all geostationary imagery of Asia was obtained from the Japanese GMS satellite.

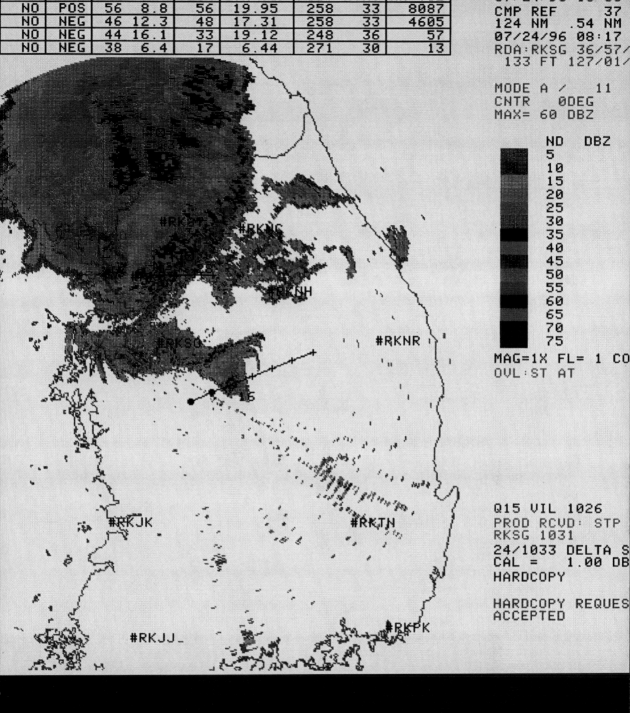

Radar

The entire radar section of this book is dedicated to the United States WSR-88D radar network. Though this admittedly does not help most international readers, the underlying principles are somewhat similar and can be used to understand the slowly emerging products from progressive radar programs such as the publically-available networks found in South Africa, Australia, Mexico, Canada, and China.

History

The roots of the Next Generation Radar (NEXRAD) Program go back to 1977. At this time, a network of WSR-74C and WSR-57 radars had been in operation for about 20 years. Research was beginning to prove the value of velocity data in storm detection. Studies began to see if the radars could be upgraded. The Joint Doppler Operations Project was established in 1979 at the National Severe Storms Laboratory by the National Weather Service (NWS), the Federal Aviation Administration, and the U.S. Air Force. The task would be to oversee the development of a next-generation radar.

During much of the 1980s, engineering was performed by Paramax, a division of Unisys, with algorithm development largely the responsibility of the National Severe Storms Laboratory. Radar units were delivered between 1990 to 1998, gradually converting the entire national radar network. In 1996 the last of the old-guard WSR-57 radars was retired in Charleston, South Carolina, marking the end of an era.

Engineering design

The WSR-88D is broken up into two main parts: the RDA and the RPG.

The RDA (Radar Data Acquisition) unit consists of the antenna, transmitter, receiver, and signal processor. This produces a raw data stream consisting of reflectivity, velocity, and spectrum width data. The RPG (Radar Products Generator) converts the raw data stream into an array of useful products, which are disseminated to all users.

Both the RDA and the RPG are controlled by a UCP (Unit Control Position) terminal, which is usually located remotely at the closest National Weather Service office. Up until the late 1990s, National Weather Service forecasters looked at the radar data using a proprietary PUP (Principal User Processor) workstation, however this has been largely replaced by WDSS (Warning Decision Support System) workstations running the Unix OS.

The power output of the WSR-88D radar is 750,000 watts. This is comparable to the transmitter of a large television station. Its klystron tube delivers a frequency of 2700 to 3000 MHz (10 to 11.1 cm; in the S-band) using a 28-foot dish that produces a beam 0.95° in width. A pulse length of 1.57 microseconds (1545 ft) is possible with the radar system. The radar collects reflectivity data at 250 m (820 ft) resolution, but this is downsized to 1 km before running detection algorithms and producing products.

The WSR-88D radome is 39 feet in diameter and made of rigid fiberglass. It is meant to protect the antenna from wind, lightning, and damaging weather. Since the dome only causes a 0.6 dB signal loss it is almost transparent to the radar unit. The radar may sit upon a tower anywhere from 20 to 98 ft in height, depending on the terrain and obstructions.

Meteorological design

The WSR-88D is designed to operate in two radically different modes: *clear air* and *precipitation*. It can only operate in one of these modes at any given time. The main difference between the two is that clear air mode offers the advantage of greater sensitivity due to a slower antenna

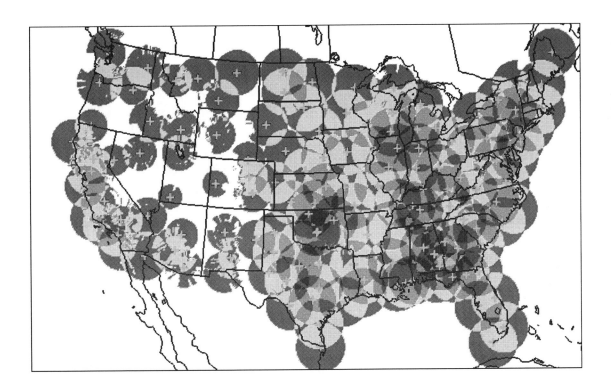

Above: Effective 128 nm radar coverage of the United States at 5 km (16400 ft) MSL, compensated for terrain and obstructions. Coverage is sparse in the Rockies, though on the other hand since the radar is usually at a high elevation in the Rockies very little of the volume is below 5 km MSL. *("WSR-88D Radar Coverages", Jian Zhang, National Severe Storms Laboratory, 2001)*

rotation rate, which allows more energy to be returned back to the radar. This comes at the cost of poor temporal resolution, with products generated half as often and sampling too slow to keep track of precipitation and especially storms.

The radar operating modes are further subdivided into a number of *volume coverage pattern* schemes, or VCPs. For a number of years there were only four VCPs: 11, 21, 31, and 32. VCP 11 was for convective precipitation, 21 for stratiform precipitation, 31 for long-pulse clear air in weak winds, and 32 for short-pulse clear air in high winds. Starting in 2003 a large number of variations of these VCPs have been introduced, all of which offer more choices for elevation, pulse length, and so forth but are based roughly on the four basic VCPs.

Another feature which can't be overlooked is the ability to provide velocity (Doppler shift) measurements of weather targets. This is accomplished by measuring the frequency shift of each bit of backscattered radiation. The velocity data allows measurement of any scatterer at any level and helps forecasters obtain wind measurements and locate circulations.

Base reflectivity

Radar energy is reflected from water droplets and ice particles back to the radar receiver. The stronger the reflection, also known as *power*, the more "intense" the echo. The WSR-88D radar detects only precipitation. It does not usually detect clouds and fog, which are primarily composed of microscopic droplets that are too small for the radar to see, except when droplets grow or ice crystals form. Base reflectivity can be contaminated by ground clutter and anomalous propagation. Insects and birds can produce large patterns of faint reflectivity around the radar site. The beam of a radar unit widens out to 2 miles in size as it gets 100 nm from the radar site, so small-scale storm features are lost at long ranges.

The WSR-88D reflectivity product provides 1 deg x 1 km resolution. In 2008, a new Super Resolution product, available at select websites and with special radar viewing software, cut this down to 0.5 deg x 0.25 km. Intensity is described in decibels of power reflected, or dBZ. The scale is logarithmic, so an increase of 3 dBZ is a doubling of power returned.

The technique of analyzing strong thunderstorms is a science in itself. Strong winds can shape the precipitation core into unique patterns, such as the hook echo, which is indicative of a tornado. A strong thunderstorm core which shifts more toward the edge of the cell, rather than remaining centered, is indicative of a severe thunderstorm. The biggest threat of severe weather is at the strong reflectivity gradient.

A feature called a "bright band" is often observed in stratiform winter precipitation situations. This is a ring that appears on the base reflectivity product, centered on the radar site. It occurs when the radar beam intersects a layer of snow melting into rain as it falls, which creates enhanced radar reflectivity. As the range and height to this feature is similar across the region, it appears at a constant range from the radar, and thus appears as a ring or a partial ring. The height of the feature can be easily estimated.

It must be remembered that the sweep of the radar beam is conical in shape; it is near the ground at the radar site and increases to higher heights at increasing distance from the radar site. Therefore echoes far from the radar are being sampled at a much higher elevation — as high as 15,000 ft at 100 miles from the radar. This makes it impossible to directly sample low-altitude features such as hook echoes at such ranges. Also a much larger volume is being sampled, so key storm features may be smeared out at these distant ranges.

a closer look

The base reflectivity product has a strong correlation to rainfall intensity, and especially to the presence of hailstones.

Most "local radar" views seen on television are base reflectivity. However anything covering an entire state or region, built up as a mosaic of multiple radars, will usually be a composite reflectivity image (q.v.). It is important to know the difference when making assumptions about weather systems based on mosaics. Quite often a composite reflectivity view can make a storm system appear much more extensive than it really is!

In clear air mode, reflectivity shows echoes spanning -28 to 28 dBZ. In precipitation mode, echoes range from 5 to 70 dBZ. Websites typically assign radically different color sets to each mode so that it is immediately obvious which color set and radar mode is in effect. For example, precipitation mode is most often made up of greens and yellows, while clear air mode shows red and gray echoes.

Always be alert for suspicious echoes, such as chaff, birds, and solar spikes. They are surprisingly common.

Radar intensity relationships:
10 dBZ - Very light rain / light snow
20 dBZ - Light rain / heavy snow
30 dBZ - Moderate rain
40 dBZ - Heavy rain / thunder
50 dBZ - Torrential rain and thunder
60 dBZ - Thunder with rain and hail
70 dBZ - Rare; usually indicates large hail.

The Product Code of base reflectivity is 19/R.

websites

weather.cod.edu/analysis/radar.main.html
www.nws.noaa.gov/radar
weatheroffice.gc.ca/radar
www.weathertap.com (pay site)
www.weathermatrix.net/radar/data/world

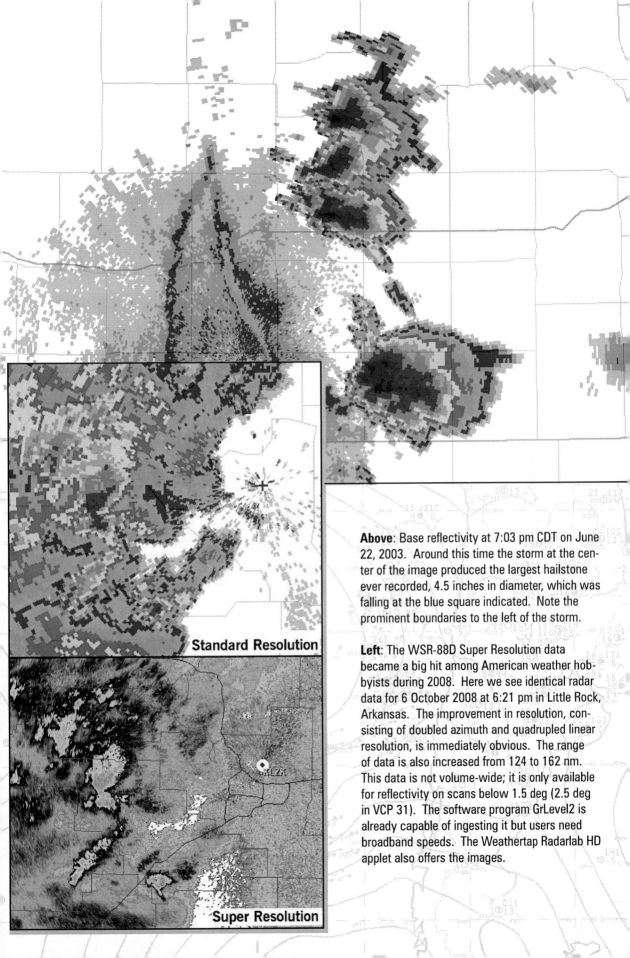

Standard Resolution

Super Resolution

Above: Base reflectivity at 7:03 pm CDT on June 22, 2003. Around this time the storm at the center of the image produced the largest hailstone ever recorded, 4.5 inches in diameter, which was falling at the blue square indicated. Note the prominent boundaries to the left of the storm.

Left: The WSR-88D Super Resolution data became a big hit among American weather hobbyists during 2008. Here we see identical radar data for 6 October 2008 at 6:21 pm in Little Rock, Arkansas. The improvement in resolution, consisting of doubled azimuth and quadrupled linear resolution, is immediately obvious. The range of data is also increased from 124 to 162 nm. This data is not volume-wide; it is only available for reflectivity on scans below 1.5 deg (2.5 deg in VCP 31). The software program GrLevel2 is already capable of ingesting it but users need broadband speeds. The Weathertap Radarlab HD applet also offers the images.

Composite reflectivity

The theory behind composite reflectivity is exactly the same as base reflectivity (see previous pages) However composite reflectivity examines not just one scan elevation but all of them. From this it displays the maximum reflectivity found in any of the "bins" vertically above a given location. In short, it displays the highest detected reflectivity value above that geographic location. This makes composite reflectivity suitable for instant display of elevated precipitation, droplets suspended high in an updraft, and even cloud decks!

One of the best uses of composite reflectivity is in detecting the first signs of thunderstorm development. The first detectable echoes within a towering cumulus transitioning to a cumulonimbus cloud will typically occur at a height of 10,000 to 20,000 ft. Within about 50 to 100 nm of the radar, this is usually *above* the lowest base reflectivity scan. Therefore the first echoes will typically show up on one of the higher base reflectivity scans and appear on the composite reflectivity product.

Since composite reflectivity merges echo information at numerous levels, important structures seen on one base reflectivity frame can be completely lost. Hook echoes, vaults, bright bands, and other interesting features will be smeared out or lost. Therefore base reflectivity should always be monitored during periods of severe weather.

Also it is not possible to draw conclusions about three-dimensional structure using composite reflectivity. Three-dimensional assumptions can be easily estimated using base reflectivity since the height is obvious to us as a function of scan height and range. For example, an intense signature close to the radar site on the 0.5° scan implies it is close to the ground. However this assumption cannot be made using composite reflectivity. Investigation of other products is necessary.

The composite reflectivity product is limited by the scan height. The radar antenna only rises to about 20 degrees, so it is incapable of seeing any tall echo near or above the radar site. A layer of elevated precipitation will appear to have a hole over the radar site as a result.

There is also a cost in terms of time. Base reflectivity products are available immediately after one full sweep of the radar beam. However, since composite reflectivity is a volumetric product, it will not be available until the entire volume scan is complete. While you are looking at a composite reflectivity product for eight minutes ago, a fresh base reflectivity scan may already be available for your use!

a closer look

Composite reflectivity is a depiction showing the highest reflectivity detected at a given location, regardless of the elevation slice being used. It uses the entire volumetric scan.

Precipitation areas look bigger than they really are at the surface. This is especially true of large thunderstorms and in dry air masses. Since many radar mosaics (maps of multiple radars) use composite reflectivity, it is important to be aware of whether you are looking at base or composite reflectivity at all times.

Always use composite reflectivity when expecting precipitation, since monitoring just one base reflectivity elevation may cause developing echoes at another elevation level to be missed. Storm echoes develop initially at the 15 to 20 thousand ft AGL height, which is above most of the 0.5 deg elevation.

Do not use composite reflectivity to examine severe storm features and boundaries, since fine detail at one particular elevation will get blurred out by other elevations. Hook echoes are severely degraded in composite reflectivity since the concave area of the hook is usually topped at higher levels by the storm's overhang.

The limitations of composite reflectivity include masking of features by echoes at other levels, inability to determine structure, scan height limitations, and time considerations since this is one of the last products generated in a volume scan. Its best use is to provide a quick overview of conditions throughout the volume scan.

The product code for Composite Reflectivity is 38/CR (for 2.2 nm 248 nm).

websites

weather.cod.edu/analysis/radar.main.html
www.nws.noaa.gov/radar
www.weathertap.com (pay site)

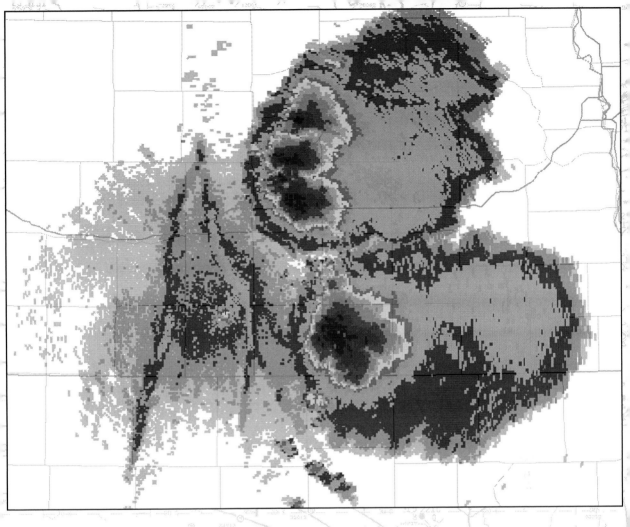

Above: Composite reflectivity for the evening of June 22, 2003, at the same time as the image in the previous section ("Reflectivity"). The low-level hooks and gradients are not apparent here due to considerable mid- and high-level reflectivity, illustrating one drawback of the composite reflectivity product. However the image does show considerable overhang of extremely high reflectivities over Aurora (blue square) where the largest hailstone on record was falling.

Right: Mosaics of multiple radars often use composite reflectivity. If using mosaics for forecasting use, it's important to determine whether the image is built from base or composite reflectivity. *(Weathertap.com)*

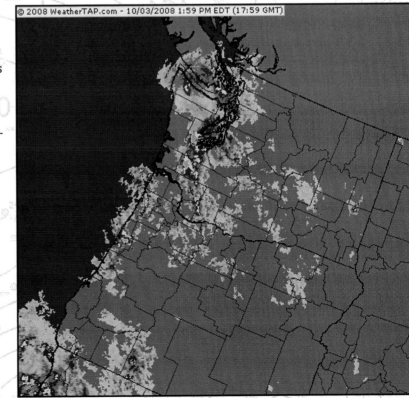

© 2008 WeatherTAP.com - 10/03/2008 1:59 PM EDT (17:59 GMT)

Velocity

In its simplest terms, the Radial Velocity product describes the radial velocity (along the beam) of scatterers within a given bin. The radar determines this by measuring the Doppler shift of the reflected energy. Since velocity is only available radially, this makes the product challenging to interpret. Tangential and pure two-dimensional motion cannot be measured; only assumed.

The overall shape of the broad-scale negative and positive shading tends to outline the tropospheric wind flow. The radar swath defines a cone within the troposphere, so all heights are included in the scan. For example a spiral appearance to the broad-scale coloring suggests change in wind direction with height. A lack of a spiral appearance suggests unidirectional winds with height.

Small-scale circulations are analyzed by finding a couplet; in other words, a pair containing positive and negative velocity. The exact orientation of this couplet relative to the radar determines whether convergence, divergence, cyclonic rotation, or anticyclonic rotation is present (see illustration). When peak velocities of a couplet touch each other, velocity is expressed in terms of "gate-to-gate shear"; this usually only occurs with the tight rotation or convergent rotation signature of a tornado.

When possible, the Storm Relative Motion (SRM) product should be used to analyze a storm. This attempts to balance velocity data by compensating for the drift of features with the prevailing flow, and can help features stand out much better. The motion is automatically derived from an average of all storm velocities as determined by the Storm Tracking Algorithm.

Range folding is an artifact that occurs when the distance to a storm exceeds the maximum unambiguous range. Beyond this range the echo arrives after the next pulse has been transmitted and a false echo occurs. Echoes that are range-folded are usually shaded gray or purple to show that they have no usable data.

Aliasing is another problem. The velocity of a scatterer may exceed the maximum unambiguous velocity (the Nyquist co-interval): the highest velocity observable by the radar given its current pulse settings. This can mask tornadic couplets by giving strange speed readings in the couplet. Dealiasing schemes do exist. Velocity products on the web are generally not dealiased. Radar viewing programs like GRLevel2 have a dealiasing toggle which should always be used in severe weather situations.

a closer look

Positive velocity is movement away from the radar. Negative velocity is movement toward the radar. These are key concepts!

There is a de-facto shading scheme in use for velocity. Positive (away) velocity usually gets a warm shade like red, while negative (toward) velocity usually gets a cool shade like green or blue. This rule of thumb can quickly help get one's bearings when the product legend is not available.

The Storm Relative Motion product should not be used by itself to make assumptions about surface winds. The SRM product uses a storm-relative frame of motion, while the Radial Velocity product uses a ground-relative frame of motion. In a downburst situation, the SRM product can help locate mechanisms for high winds, while the radial velocity product suggests the actual winds experienced at the surface (relative to the radar, of course).

If you trace the border formed between the broad-scale negative and positive areas on the velocity scan, it can be used to judge the wind direction throughout the entire troposphere. At any given point along this line, the wind is perpendicular to the radar beam, blowing toward the positive velocity area. By tracing this line from the radar site to the outer range area, you can follow the wind direction from the ground to the upper troposphere or stratosphere. The wind speed is traced in a like manner by going outward from the radar site to the maximum range and finding the maximum wind velocity at each particular range.

Product Code for radial velocity is 27/V (25/V for 32 nm/0.13 nm)

websites

weather.cod.edu/analysis/radar.main.html
www.nws.noaa.gov/radar
www.weathertap.com (pay site)

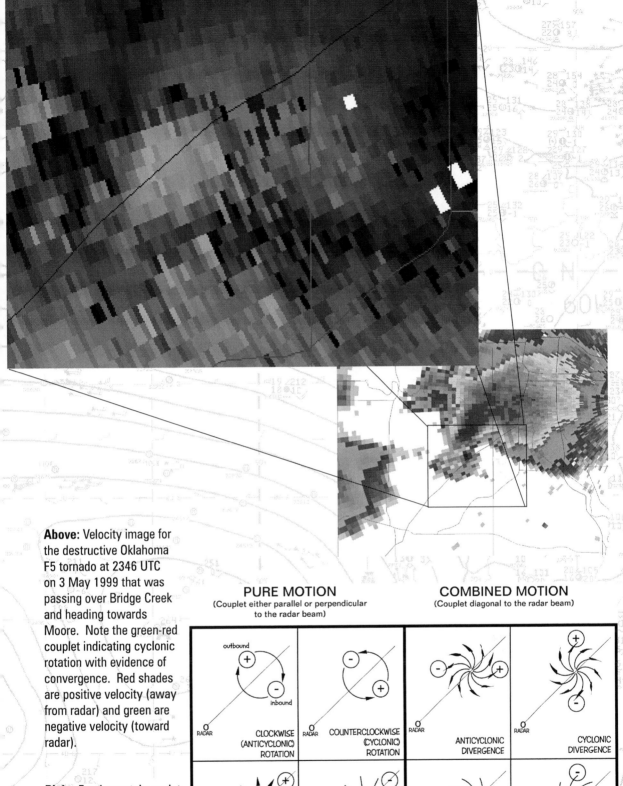

Above: Velocity image for the destructive Oklahoma F5 tornado at 2346 UTC on 3 May 1999 that was passing over Bridge Creek and heading towards Moore. Note the green-red couplet indicating cyclonic rotation with evidence of convergence. Red shades are positive velocity (away from radar) and green are negative velocity (toward radar).

Right: Fundamental couplet models for different types of small-scale circulations. The orientation of the couplets relative to the radar determines what type of circulation is present.

PURE MOTION
(Couplet either parallel or perpendicular to the radar beam)

COMBINED MOTION
(Couplet diagonal to the radar beam)

outbound

inbound

O
RADAR CLOCKWISE
(ANTICYCLONIC)
ROTATION

O
RADAR COUNTERCLOCKWISE
(CYCLONIC)
ROTATION

O
RADAR ANTICYCLONIC
DIVERGENCE

O
RADAR CYCLONIC
DIVERGENCE

O
RADAR PURE DIVERGENCE

O
RADAR PURE CONVERGENCE

O
RADAR ANTICYCLONIC
CONVERGENCE

O
RADAR CYCLONIC
CONVERGENCE

Spectrum width

Spectrum width measures the variance in velocity within a given radar volume. Particles in any given volume are almost never moving cohesively at a single velocity. Turbulence, varying particle sizes, and different forces at work within the volume can all combine to cause varying trajectories and velocities. The velocity estimated by the radar in one specific bin is actually an average of all of these motions.

The variance in the motion corresponds to the spectrum width. This can be measured by examining the "width" of the Doppler shift that is returned to the radar. Low width will cause the reflected energy to peak at one very specific frequency, while high width will cause a broad dispersion of the echo at different frequencies. A spectrum width is expressed as low (narrow) or high (broad).

Unfortunately spectrum width has been largely neglected in the study of operational meteorology. One of the more interesting investigations into spectrum width was a paper by Keith Browning, drawn upon by Leslie Lemon. It suggested that an updraft core normally contains air that ascends uniformly and smoothly, with turbulent flow largely dampened out. This implies that the updraft may be found by locating a core of narrow spectrum width within the cloud.

There is some evidence that spectrum width products can help locate small tornadoes, gustnadoes, and waterspouts. Within a radar bin, a small tornado may not contain enough fast-moving scatterers to trigger a shear signature. However these scatterers would produce a wide range of velocities within that particular bin. The spectrum width product would return a high value. Spectrum width can help locate and measure areas of turbulence.

Another use of spectrum width is to identify suspected three-body scatter spike (TBSS) signatures, which are artifacts that seem to project behind a hail core along the beam radial. The presence of a high spectrum width can help suggest that the artifact is indeed a TBSS artifact.

In more quiescent weather patterns, spectrum width data can be used to help locate fronts and outflow boundaries. Often these features are easily identified, particularly in clear air mode, but at times they may be masked. Spectrum width can help confirm the validity of radial velocity data. A large spectrum width may indicate that the averaged radial velocity for that bin is not reliable. Finally, spectrum width may also help locate initial convective development.

a closer look

Spectrum width gives information on the variance of velocity within a particular radar bin. It is an indirect measure of turbulence and gustiness as expressed by the motion of rain, ice, and other particles.

Spectrum width tends to increase with range from the radar. This is a natural consequence of beam broadening: as the beam widens, there is bound to be an increasingly large range of particle motions.

Spectrum width values are not affected by VCP changes, though clear-air modes provide better sampling and more accuracy.

Spectrum width is one of the three base products of the WSR-88D, in addition to reflectivity and velocity.

The idea of using spectrum width to identify small vortices was pioneered in 1995 by Joseph Golden and Carin Goodall-Gosnell, and in another 1995 paper by Waylon Collins. Both papers studied waterspout events in the southeastern United States and related them to WSR-88D spectrum width data.

WSR-88D spectrum width data has been the focus of several studies by UCAR and other institutions designed to help reduce the risk of clear air turbulence to aircraft.

Weak reflectivity bordering on the noise threshold for the radar will cause erratic spectrum estimates and noisy spectrum width returns.

The product code is 30/SW (124/0.54 nm) and 28/SW (32/0.13 nm).

websites

None are known to exist! This information can be plotted with standalone software like GRLevelX and Digital Atmosphere.

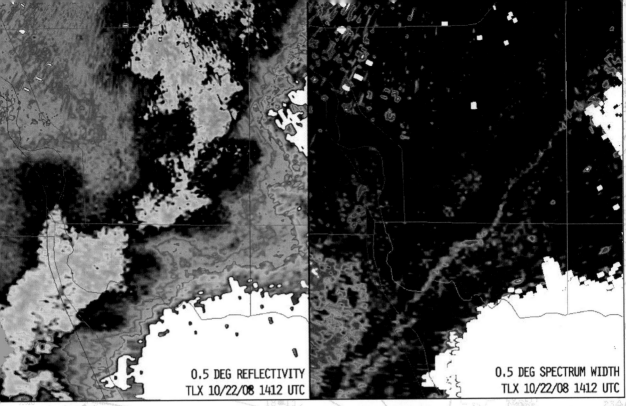

0.5 DEG REFLECTIVITY
TLX 10/22/08 1412 UTC

0.5 DEG SPECTRUM WIDTH
TLX 10/22/08 1412 UTC

Above: Super-resolution WSR-88D imagery along a line of storms in southern Cleveland County, Oklahoma showing reflectivity (left) and spectrum width (right). Note the snake-like outflow boundary clearly displayed stretching from northeast to southwest along the leading edge of the storm line.

Below: WSR-88D spectrum width product from the Denver WSR-88D. *(UCAR)*

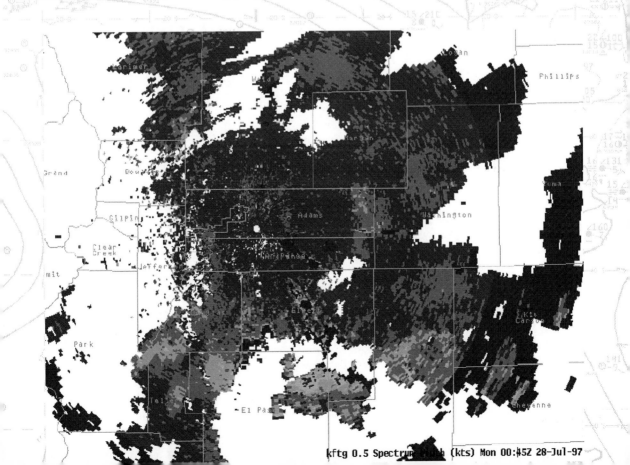

kftg 0.5 Spectrum Width (kts) Mon 00:45Z 28-Jul-97

Precipitation total

The Precipitation Total product actually refers to three specific products generated by the NEXRAD site. These are 1-hour precipitation, 3-hour precipitation, and storm total precipitation. These provide a graphic estimate of precipitation totals within 124 nm of the radar site. The data is expressed as 16 levels from 0 to 15 inches.

The Precipitation Total product is excellent for monitoring areas that have had excessive precipitation. These areas may be subject to flooding. It can also reveal where grounds are saturated, which can further compound flooding and can feed moisture back into a convective situation through evaporation. The product can determine where heavy snow has fallen. Hydrologists will use Precipitation Totals to find where basins are approaching saturation and where streams might reach flood stage.

If no precipitation, as determined by the precipitation totals algorithm, has fallen within 124 nm of the site after more than an hour, the storm total precipitation (STP) product is automatically reset and all plots show a zero precipitation total. However, during long rain events the storm total period may exceed 24 hours.

As with many other products, precipitation total estimates are limited by the tilt of the radar antenna, which reaches no higher than 19.5°. Not all of a precipitation shaft which is very close to the antenna will be sampled. The problem usually occurs within 20 nm of the site for convective precipitation and within 10 nm of the site for stratiform precipitation.

The data is easily contaminated by ground clutter and anomalous propagation. Chaff (from military aircraft) will produce false totals and produce plots that look unusually like precipitation. When bright bands occur during cold weather events, these will distort the totals.

Fast-moving systems will distort the precipitation total pattern, as the storm will move across multiple bins between scans. This will produce a herringbone pattern in the precipitation total "trails" left by storms.

Hail will cause overestimation of rainfall amounts, as hail particles reflect much more power back to the radar than water droplets do. The three-body scattering spike signatures can also "lay down" precipitation values behind the hail core, where no precipitation actually exists.

It should be noted that there is no quality control of the precipitation product at any stage. Users have to evaluate the results against their knowledge of radar principles, climatology, and empirical knowledge of the precipitation total accuracy.

a closer look

The depiction of precipitation totals was changed from Cartesian 1.1 nm blocks to standard polar coordinate format effective with NEXRAD Build 9 in late 1996.

Precipitation total estimates may be degraded by the presence of ice, snow, and especially hail. The algorithm works best with warm, slow-moving precipitation.

It is useful to conduct a study of your local area and determine whether the radar tends to underestimate or overestimate precipitation. Be sure to use official day-to-day readings at an airport weather station for representative results, rather than using extreme rainfall amounts that could be difficult for the radar to estimate.

Storm total precipitation values are reset when no precipitation totals are detected by the algorithm for at least one full hour.

Storm total precipitation is not confined to a specific time limit.

Plots near military operating areas are frequently degraded by chaff released by military aircraft during mock dogfights. Notorious chaff regions are the area north of Las Vegas, the area west and southwest of Phoenix, and the region southwest of Salt Lake City.

Precipitation total estimates are corrupted by precipitation too close to the radar antenna, by ground clutter or chaff, by fast moving storms, and by hail cores.

The Product Codes for precipitation products are 78/OHP (1-hr); 79/THP (3-hr); and 80/STP (storm total).

websites

weather.cod.edu/analysis/radar.main.html
www.nws.noaa.gov/radar
www.weathertap.com (pay site)

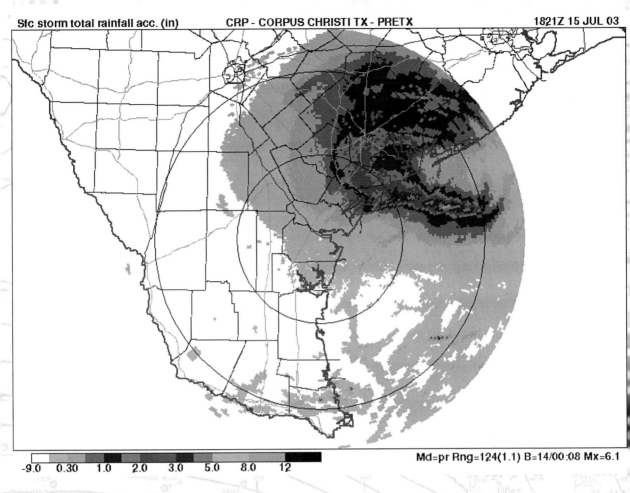

Sfc storm total rainfall acc. (in)   CRP - CORPUS CHRISTI TX - PRETX   1821Z 15 JUL 03

| -9.0 | 0.30 | 1.0 | 2.0 | 3.0 | 5.0 | 8.0 | 12 |

Md=pr Rng=124(1.1) B=14/00:08 Mx=6.1

Above: Storm total precipitation product generated during the landfall of Hurricane Claudette on July 15, 2003. *(College of DuPage)*

Right: A National Weather Service hydrologist adjusts a 1-hour precipitation total estimate from the Fort Worth WSR-88D before feeding it into a local model. This quality control procedure makes for better runoff and soil saturation estimates, improving the quality of flood forecasts. *(Tim Vasquez)*

Vertically integrated liquid

Vertically Integrated Liquid, or VIL, indicates the average mass of liquid water per cubic meter within a column above the earth's surface. It examines all of the 2.2 x 2.2 nm radar bins above a given location and adds them to estimate the total water content above a given location. The estimation of liquid precipitation within each bin uses complex equations that relate reflectivity to drop-size distribution and water content. VIL is initially calculated in terms of kilograms of water per square meter, then is integrated throughout the column to produce a volumetric measure in kilograms per cubic meter.

The most important strength of VIL is that, being an additive total of all echoes in the vertical, it allows an immediate assessment of which storms are most important. A cyclic increase and decrease in VIL values indicates a multicell thunderstorm structure, though this could also indicate a storm that is not being adequately sampled in the upper levels (especially in VCP 21). Persistent high VIL values are usually associated with supercells.

Furthermore, VIL has formed one of the strongest links to observed hail size. In fact, during the early 1990's forecasters were urged to use VIL to assess the hail threat. The problem with this approach, however, is that the VIL algorithm filters out extremely high reflectivity, associated with hail, in order to get an accurate VIL quantity. The introduction of a new Hail Detection Algorithm, which properly links the melting level to significant storm tops and to the surface, has allowed for better indicator of hail.

VIL is affected by any process or phenomena that distorts reflectivity values. Therefore, chaff, bright bands, and three-body scatter spikes can all corrupt a VIL value.

A storm that is tilted will produce artificially low VIL values. A tilted storm spreads its footprint across multiple vertical bins, thinning out the VIL response.

A fast moving storm will move from one horizontal grid square to the next between each elevation scan. When the volume scan is complete, the storm may have crossed ten grid squares or more (i.e., a storm moving 40 mph will move 4 miles during a volume scan). This will distort the VIL calculations and produce a lower VIL value spread across a wider area.

Once the storm becomes 40 nm or further from the radar, gaps between the elevation scans become increasingly wider and more of the storm becomes unsampled. The bins also become very wide, and intense cores may not be fully detected. Therefore VIL values tend to be overestimated beyond about 110 nm.

a closer look

VIL is an indirect estimate of the instantaneous liquid content above a given spot. It can be thought of as reflectivity summed with height. VIL was often used in the 1990s to find cells with a high hail threat, however the hail product now considered much better for this purpose.

VIL is closely correlated to updraft strength and can be used to quickly identify cells that require monitoring for severe weather or flooding.

There is no magic number for VIL values. They vary according to the season and type of weather system. Trends and immediate differences are more meaningful.

VIL values are the least reliable in VCP 21 due to the wide elevation gaps present above 4.3°.

VIL is adversely affected by storm tilt, fast storm motion, elevation gaps, proximity to the radar, and operation in VCP 21.

VIL will not be adequately represented closer than about 20 nm to the radar, as the radar antenna is limited to a tilt of 19.5°.

You may encounter the term "VIL density". This refers to VIL divided by the echo top (it is usually expressed in units of 1000). So a storm with a low top would have a high VIL density, given the same VIL value. The idea is that it is normal for shallow storms to have low VIL values and for deep storms to have high VIL values, and to help identify any departure from this basic relationship.

The Product Code of VIL is 57/VIL.

websites

weather.cod.edu/analysis/radar.main.html
www.nws.noaa.gov/radar
www.weathertap.com (pay site)

Right: VIL values for storms in Nebraska on June 22, 2003, same as the storms shown on the "Reflectivity" and "Composite Reflectivity". The small red square on the center storm represents Aurora, where the record hailstone fell. The white blocks at the centers of the storms indicate extreme VIL values exceeding 70 kg/m³, which are usually associated with hail but may also indicate deep storms and clouds with an excessive load of large rain drops. *(Generated with Digital Atmosphere)*

Right: A cross-section of a storm shows the simplified concept of VIL. Arbitrary reflectivity values are shown within the storm cloud (outlined). The values are summed to produce a total. Notice how the lowest reflectivity scans miss any significant precipitation, but VIL detects the higher core. Composite reflectivity is similar in principle and would show the highest value in each vertical stack.

The basic equation for VIL is:
$\text{VIL} = \text{SUM } 3.44 \times 10^{-6} Z^{4/7} \delta h$
where Z is the average radar reflectivity within a layer and δh is the thickness of that layer.

Echo tops

The Echo Tops product is a processed type of image. The region is divided up into grid boxes measuring 2.2 nm square, and all scans are analyzed to find the highest grid box that exceeds 18.5 dBZ. The center beam height for the highest scan meeting the reflectivity threshold is the echo top.

The biggest problem with this product is that the echo tops are not always sampled, or may be sampled too coarsely.

The first type of problem is related to the antenna's maximum tilt, which is 19.5°. As a result, when storms exceed 30,000 ft within 15 nm of the radar or 50,000 ft within 25 nm of the radar, the radar is unable to scan clear air above the storm and the echo top is underestimated.

Elevation gaps are a significant problem, in which are slices "skipped" by the scan strategy. This is a significant problem in VCP 21 which has only four scans between 4.3° and 19.5°. For example, imagine a storm with a top of 55,000 ft located 55 nm from the radar. The 6° elevation intersects this range at 60,000 ft and the 4.3° elevation intersects at 37,000 ft. Unfortunately it is only the 4.3° elevation that detects the storm, so the elevation is counted as 37,000 ft. As an echo gets closer, its elevation seems to drop as it descends the beam until it is detected on the next highest elevation slice and jumps to another height. This behavior causes the "stairstep" pattern often seen on Echo Tops products. It is most prominent when stratiform tops are present.

Beam width is also a detrimental factor. At 120 nm the width of the radar beam spreads to about 13,000 ft. This means that a storm registering a top of 48,000 ft top may actually be topping out at anywhere between 38,500 and 51,500 ft.

All of this may be further compounded by the limited resolution of the Echo Tops product. Storms are rounded into blocks of 5,000 ft. As a result, a storm with a height of 29,000 ft will fall into the 25,000 ft to 30,000 ft block, and may be interpreted to be slightly weaker than it really is.

The threshold value of 18.5 dBZ means that an echo top with a reflectivity of less than 18.5 dBZ will be cut off. An echo top, which includes an overshoot or anvil top, may be actually higher than what is depicted.

Finally, the echo top product does not compensate for sidelobes, which are artificial spikes that are generated due to the extreme reflectivity near hail cores. This may lead to the product *overestimating* echo tops.

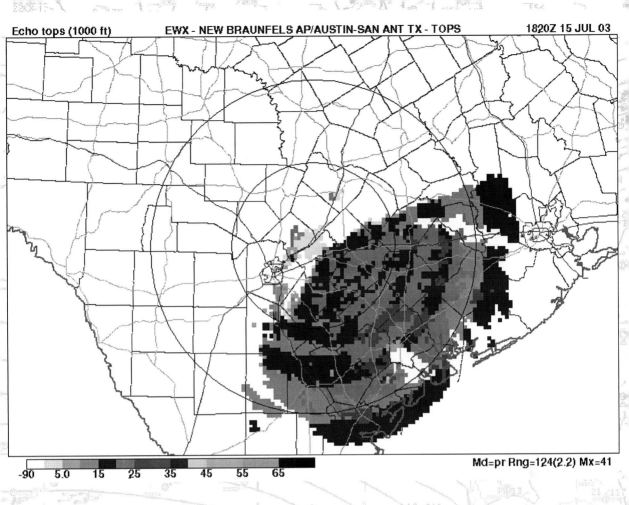

-90   5.0   15   25   35   45   55   65

Md=pr Rng=124(2.2) Mx=41

Above: Echo tops product generated during the landfall of Hurricane Claudette on July 15, 2003. Note the stadium appearance of the hurricane due to sampling of the top at different elevations. *(College of DuPage)*

Right: This diagram demonstrates the problems caused by elevation gaps as a storm draws closer, using three thin beams for emphasis to judge each cloud top. As a result, the storm seems to get shorter, then jump to 40,000 ft as it approaches. This produces a stadium effect on echo top products. The sampling problem is worst in VCP 21.

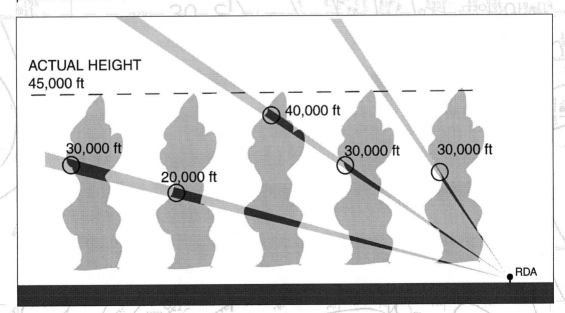

ACTUAL HEIGHT
45,000 ft

40,000 ft

30,000 ft

20,000 ft

30,000 ft   30,000 ft

RDA

Storm tracking information

The purpose of the Storm Tracking Information product is to identify storm centroids and display their past, current, and projected locations. It uses the Storm Cell Identification and Tracking (SCIT) Algorithm, which replaced the old Storm Series Algorithm that was phased out with NEXRAD Build 9 in late 1996.

First the Storm Segments algorithm is run. One-dimensional storm segments are located by searching along a radial (i.e. at different azimuths at a constant distance) for one-dimensional runs of contiguous high reflectivities. If a series of high reflectivities has suffcient radial length to meet or exceed a parameter called the "overlap threshold", it is classified as a storm segment.

The Storm Centroids algorithm then runs. It gathers all storm segments and attempts to build two-dimensional representations of each storm. This produces a storm component. Each storm component is checked to see if it meets the minimum accepted size. Generally a size of 2.2 nm is considered sufficient. If it does not meets the size criteria, it is deleted. Each storm component is measured and assigned a "disk" representing its location and radius.

When this is complete, storm components are resolved into three dimensions by looking at all elevations, measuring overlap of components, and determining an overall centroid for the storm.

At this point, the Storm Tracking Information algorithm executes. The algorithm's job is to relate each storm found in a current volume scan to a storm found in a previous volume scan. The algorithm starts with the largest storm centroid and works down to the smallest ones, linking current storm centroids to those from previous volume scans. This builds a track history. There are also sanity checks that are performed. A storm motion change over time must not exceed 90 degrees from the last scan. A storm centroid may also not have changed mass by more than one order of magnitude.

Finally the Storm Position Forecast algorithm is run. Each storm's anticipated motion during the upcoming hour is made by a simple extrapolation of its average direction and speed of movement. The algorithm also checks its work; if it has performed poorly, no forecasted positions are displayed. However if it has done a good job, forecast centroids for the next four volume scans are displayed. No projection is made for new storms.

Storm tracking is available on some pay websites and with special software. It is important to avoid the trap of overdependence on this product, however.

a closer look

The Storm Cell Identification and Tracking (SCIT) algorithm identifies thunderstorm cells and plots a history of paths and a projected track for each detected cell in the volume scan.

Since the reflectivity product only examines precipitation cores, this product cannot be used to make interpretations about the updraft core, mesocyclones, or tornadoes.

The SCIT algorithm was formerly known as the Storm Series Algorithm. This older algorithm focused exclusively on cells that were 30 dBZ or greater and lacked a sophisticated identification scheme.

Unusual structures, especially linear configurations seen in squall lines, will cause serious problems with the SCIT algorithm and make much of the output unusable.

Forecast tracks are based only on previous movement. The algorithm does not forecast changes in intensity, path, or speed, and will not predict deviant movement.

A history track that shows a curve deviating from the mean tropospheric flow or from other cells may signal that the storm has transitioned to a severe phase.

Graphics display workstations such as WDSS often have a product available called "Cell Trends". This shows, for a given cell, its cell top, cell base, height of the storm centroid, its maximum reflectivity, probability of hail, probability of severe hail, cell-based VIL, and maximum reflectivity.

The product code is 58/STI.

websites

www.weathertap.com (pay site)

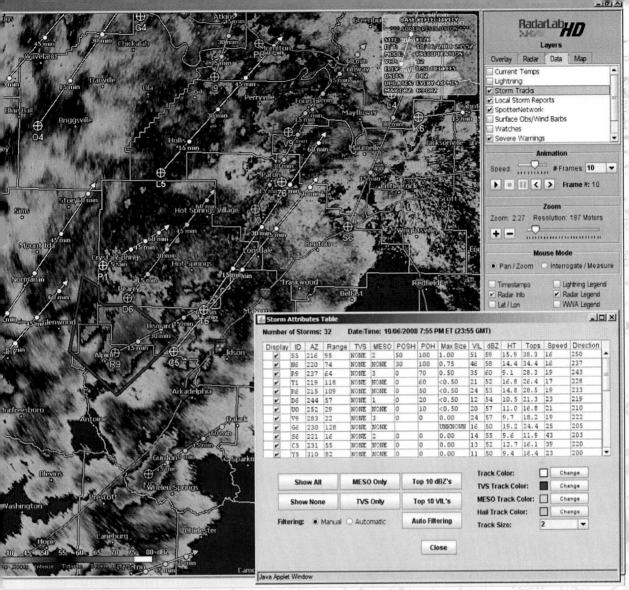

Above: Weathertap offers a fairly sophisticated Java applet that includes storm tracking information with super resolution radar. It is part of their RadarLab HD suite. Only future tracks are displayable. *(Weathertap)*

Below: Storm attributes as displayed in the standalone radar application GrLevel3.

Hail algorithm

The Hail Index is designed to display whether a storm's structure is conducive to hail formation. It incorporates the new Hail Detection Algorithm (HDA), which replaces the old Hail Index Algorithm that was phased out with NEXRAD Build 9 in late 1996. The new algorithm provides more robust detection capability, along with a new set of statistics for the end user such as hail size. An upgrade in 1998 allowed the generation of hail probability. Its input is storm centroids and components from the Storm Cell Identification and Tracking (SCIT) algorithm, along with user-defined melting level information.

The HDA produces three quantities for each location where hail is flagged: Probability Of Hail (POH), Probability Of Severe Hail (POSH), and Maximum Expected Hail Size (MEHS). In all instances, the difference in elevation between the melting level and the radar are factored in, so results are valid for the radar's elevation. Storms raking higher terrain may require special evaluation.

The Probability Of Hail parameter determines the chance of any hail of any size reaching the Earth's surface. It looks for the height of the 45 dBZ echo above the melting level. According to this algorithm, if the 45 dBZ echo is 2 km above the melting level the POH is about 20%, and if it exceeds 6 km the POH rises to 100%.

The Probability Of Severe Hail runs a process called the Severe Hail Index (SHI). This process uses storm components from the SCIT algorithm as input "objects" and relates reflectivities to melting level data. It outputs a SHI value in joules per meter per second. From this, a POSH probability value is determined.

The Maximum Expected Hail Size uses only the SHI data, and uses a simple function that increases the expected hail size as the severe hail index grows. It is considered to be the most difficult part of the forecasts output by the HDA, since it cannot forecast unusual hail shapes and minor elevation differences.

How does the algorithm perform? So far it has received excellent reviews. A study done at NWS Wichita revealed that the HDA tends to overestimate hail size, but works very well in strong mid-level flow scenarios.

It must also be pointed out that the hail algorithm may fail in unusual structures, which include highly sheared storms and deviant movers. It may also fail in squall lines, since the Storm Tracking algorithms have trouble with linear structures. Furthermore the algorithm needs to be able to measure the full depth of a storm for best results; not good for storms right over a radar site.

a closer look

The Hail Detection Algorithm provides a robust technique for measuring the hail potential of individual storm cells, calculating its probability of producing large hail and hail in general, along with maximum hail size. Hail size is an indirect indicator of updraft strength.

By convention, the symbol for hail is a triangle that points upward. Probable hail is identified by a hollow triangle. Positive hail is identified by a filled triangle.

The original NEXRAD hail detection algorithm (HDA) was based on Leslie Lemon's 1978 storm structure concept and defined by Pio Petrocchi, John Smart, and Ron Alberty in 1982-85.

The old HDA was based on seven weighted questions, in order of increasing importance: does the highest storm component reach 8 km or more; does the storm's maximum reflectivity exceed 55 dBZ; is the low-level storm component north of one at a higher level; does the storm exhibit tilt; is the mid-level reflectivity (5 to 12 km) greater than 50 dBZ; does overhang of more than 4 km exist; and does the highest storm component exist above an overhang. The resulting score was flagged as either probable or positive.

The new Hail Detection Algorithm was developed in the mid-1990s and is described in the body text of this section.

The MEHS is calculated by the HDA as:
$$MEHS = 2.54 \times SHI^{0.5}$$
where MEHS is in millimeters and SHI is the severe hail index in joules per meter per second.

The product code is 59/HI.

websites

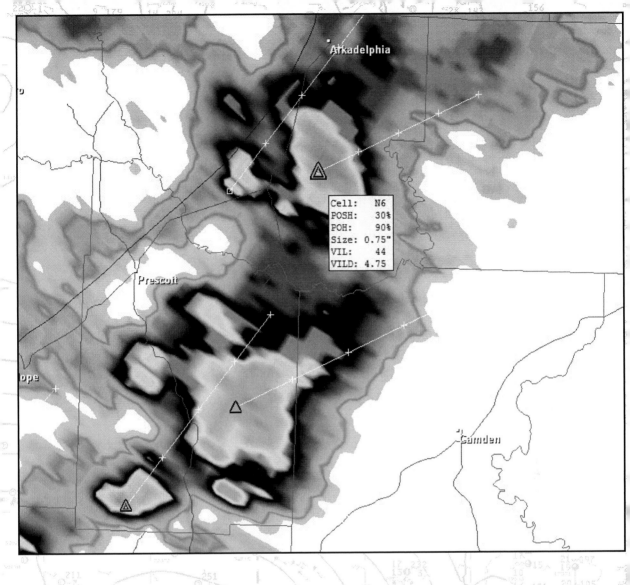

```
Cell:   N6
POSH:   30%
POH:    90%
Size:   0.75"
VIL:    44
VILD:   4.75
```

Above: GrLevel3 has a fairly robust hail plotting system. Hovering the mouse over the cell shows POSH, POH, and MEHS attributes. Storm tracks (not hail tracks!) are seen in white. *(NSSL)*

Right: Hailstone collected during Project Vortex. The oblate shape of the hailstone reveals some of the problems with trying to determine an exact forecast of hail size. *(NSSL photo)*

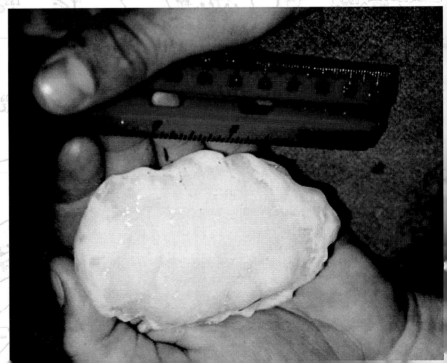

Mesocyclone algorithm

The NEXRAD Mesocyclone Detection Algorithm (MDA) pairs up base velocity information with processing power to determine the location of thunderstorm mesocyclones. The idea is not to take the responsibility away from the forecaster, but to provide the means to identify a potentially dangerous storm that might otherwise go unnoticed. The algorithm also identifies areas of uncorrelated shear which alerts the forecaster to rotation in a suspect area that is below the physical thresholds for a mesocyclone.

A mesocyclone is a region of strong, consistent rotation within a thunderstorm. It typically measures several miles in diameter and may carry tangential winds of over 40 mph. A mesocyclone signifies a very strong, organized updraft, and is often a precursor to tornado development. The vast majority of tornadoes are spawned within a mesocyclone, and perhaps all supercells contain a mesocyclone. Mesocyclones are typically cyclonic but a few may be anticyclonic. They usually develop at mid-levels and descend to the surface with time.

The processing algorithm starts from scratch with the base velocity product. At each elevation, and range, it examines all of the bins throughout the entire azimuth sweep to find groups of adjacent bins that show an increase or decrease in velocity from one end to the other. The change must span a large enough number of bins to be counted. This value is called the "pattern vector threshold" and can be modified by the radar operator. If the change exceeds the pattern vector threshold, typically ten bins, and meets a threshold value of shear and momentum, the change is classified as a pattern vector.

The MDA tries to link all pattern vectors to form two-dimensional features. When this is completed, it attempts to resolve these features into three-dimensional circulations by linking the features to others at higher elevations.

Based on the success of this three-dimensional correlation, the feature is then identified as one of three things: three-dimensional uncorrelated shear, uncorrelated shear, or a mesocyclone. Three-dimensional uncorrelated shear has vertical but not horizontal consistency. Uncorrelated shear is the opposite; it has horizontal consistency but none in the vertical. A mesocyclone, however, contains both. Finally the MDA creates an attribute table showing information about it.

The algorithm was initially developed for powerful Great Plains supercells, however a series of refinements in the mid-1990s allowed better performance nationwide.

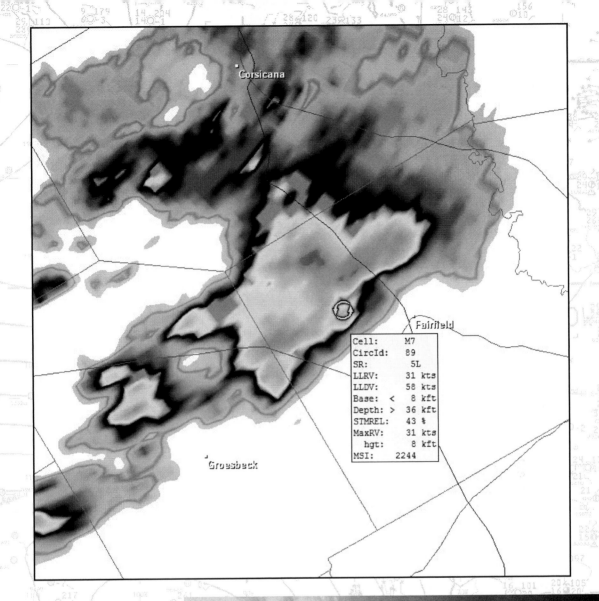

Cell:	M7
CircId:	89
SR:	5L
LLRV:	31 kts
LLDV:	58 kts
Base: <	8 kft
Depth: >	36 kft
STMREL:	43 %
MaxRV:	31 kts
hgt:	8 kft
MSI:	2244

Above: Mesocyclone plot in the GrLevel3 software program. Hovering the mouse over the mesocyclone brings up computed information as shown here.

Right: Supercell with a powerful mid-level mesocyclone near Anadarko OK on 3 May 1999. The mesocyclone is embedded within the updraft column shown here. Mesocyclone products can help increase leadtime for tornado warnings. *(Tim Vasquez)*

Tornadic vortex signature

When NEXRAD units were first fielded, they were equipped with the Tornadic Vortex Signature (TVS) algorithm. The algorithm was developed by the National Severe Storms Laboratories in the mid-1980s. It was dependent on the Mesocyclone Algorithm and looked for velocity couplets in identified mesocyclones which exceeded a threshold value. Unfortunately it was not very flexible and often missed weaker tornadoes.

The Tornado Detection Algorithm (TDA) was developed by the National Severe Storms Laboratory as a replacement. It was included in Build 10 of NEXRAD, which was released in November 1998. The TDA came with its own logic to scrutinize the entire radial velocity product, freeing it from the dependence on the Meso-cyclone Algorithm. Although the new TVS algorithm is called the TDA, the end-user product is still called "TVS" to this day.

Much like the Mesocyclone Detection Algorithm, the TDA builds one-dimensional pattern vectors, then two-dimensional features, then three-dimensional circulations. It then uses altitude, depth, and shear criteria to identify possible tornadoes. TDA actually examines the pattern vectors to calculate shear, rather than finding couplets. At least three features are required with a depth of at least 1.5 km, must have a base at 0.5 deg or below 600 meters, whichever is lower, and must have a low-level velocity differential of 49 kt or a maximum velocity differential of 70 kt.

An Elevated TVS (ETVS) is the same but may be above 0.5 deg and 600 meters and requires only a low-level velocity differential of 49 kt. It generally indicates sharp rotation aloft and ideally will alert forecasters to a developing tornado.

When scrutinizing a TDA TVS, consider whether the atmosphere is capable of providing the instability and shear neccessary for a tornado. False alarms do occur. Also note that at ranges beyond 35 nm the TDA may be triggered by strong mesocyclones.

In closing, it must be emphasized that the TDA product, as well as other algorithms, are not the holy grail of tornado detection. The forecaster must interrogate all available data, including radar images and spotter reports, and produce a coherent picture of what is happening. This must be weighed against one's own experience and in context with sound meteorological principles. Only then can a tornado forecast be considered robust and impeccable.

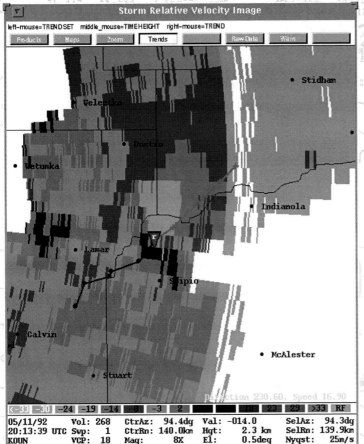

Storm Relative Velocity Image

Above: Prototype display of the TVS algorithm at NSSL shows a confirmed tornado. *(NSSL)*

Below: Almost never seen operationally, this is an actual TDA message from which TVS plots originate on radar workstations. It reveals the full scope of information available from TDA processing code. At the moment of this example output, a violent tornado was destroying the Bridge Creek, Oklahoma subdivision and was 9 miles from southwest Moore, which would be levelled. The bottom two thirds of the messages show adaptation parameters, which are threshold values set at the radar site.

```
SDUS54 KTLX 032356
NTVTLX


TORNADO VORTEX SIGNATURE
P   RADAR ID   1   DATE/TIME   05:03/99/23:56:21   NUMBER OF TVS/ETVS   3/ 0
P
P FEAT  STORM   AZ/RAN   AVGDV  LLDV   MXDV/HGT   DEPTH   BASE/TOP   MXSHR/HGT
P TYPE   ID    (DEG,NM)  (KT)   (KT)   (KT,KFT)   (KFT)    (KFT)     (E-3/S,KFT)
P
P TVS    N1    256/ 21    71    140    140/ 1.3   >29.2  < 1.3/ 30.5   108/ 3.5
P TVS    H1    305/ 42    37     94     94/ 3.3   >13.6  < 3.3/ 16.9    36/ 3.3
P TVS    E0    253/ 53    43     27    121/20.0   >25.3  < 4.6/ 29.9    38/20.0
YY
P               TORNADO VORTEX SIGNATURE ADAPTATION PARAMETERS
P
P     0(DBZ).MIN REFLECTIVITY             2.5(KM)..CIRCULATION RADIUS #1
P    11(M/S).VECTOR VELOCITY DIFFERENCE   4.0(KM)..CIRCULATION RADIUS #2
P   100(KM)..MAX PATTERN VECTOR RANGE     80(KM)..CIRCULATION RADIUS RANGE
P  10.0(KM)..MAX PATTERN VECTOR HEIGHT    600......MAX # OF 2D FEATURES
P 2500......MAX # OF PATTERN VECTORS      3......MIN # OF 2D FEAT/3D FEATURE
P    11(M/S).DIFFERENTIAL VELOCITY #1     1.5(KM)..MIN 3D FEATURE DEPTH
P    15(M/S).DIFFERENTIAL VELOCITY #2     25(M/S).MIN 3D FEAT LOW-LVL DELTA VEL
P    20(M/S).DIFFERENTIAL VELOCITY #3     36(M/S).MIN TVS DELTA VELOCITY
P    25(M/S).DIFFERENTIAL VELOCITY #4     35......MAX # OF 3D FEATURES
P    30(M/S).DIFFERENTIAL VELOCITY #5     15......MAX # OF TVSS
P    35(M/S).DIFFERENTIAL VELOCITY #6     0......MAX # OF ELEVATED TVSS
P     3......MIN # OF VECTORS/2D FEATURE   0.6(KM)..MIN TVS BASE HEIGHT
P   0.5(KM)..2D VECTOR RADIAL DISTANCE    1.0(DEG).MIN TVS ELEVATION
P   1.5(DEG).2D VECTOR AZIMUTHAL DIST     3.0(KM)..MIN AVG DELTA VELOCITY HGT
P   4.0(KM/KM).2D FEATURE ASPECT RATIO    20.0(KM)..MAX STORM ASSOCIATION DIST
YY
```

VAD wind profile

The Velocity Azimuth Display (VAD) Wind Profile (VWP) displays a time-height diagram of wind direction and speed. The data is derived from base velocity data. VWP data is provided every 1,000 ft from 1,000 to 70,000 ft MSL. The data is color coded to show the root mean square error in knots, which is inversely proportional to the reliability of the data at that height.

The VAD algorithm first waits for all reflectivity and velocity data to be processed, dealiased, and range unfolded. It then sequentially figures a wind speed at each height level. It does this by finding the scan elevation and range closest to a distance called the VAD Analysis Range or VAD Optimum Slant Range (usually 18.6 nm), which is a direct distance to the scatterer rather than a horizontal range. The algorithm analyzes all of the scatterers at that range and elevation on a sweep of all azimuths.

If 25 scatterers are detected, the algorithm attempts to fit a sine wave to the azimuth and velocity of all the data points. The sine wave configuration is used because an ideal plot of winds that are constant throughout the scan volume will resemble a sine wave if the velocity is plotted with respect to azimuth. Therefore the data should fit this curve.

The processor checks the fit of the data to the sine wave by computing a root mean square (RMS) error. This determines the reliability of the entire sample. A symmetry analysis is then done by comparing the departure of the fitted curve's baseline (zero-velocity line) from that of a standard sine wave. If the data fails the analysis by not meeting either check it is discarded and the data is considered void at that level.

During convective weather situations, changes in the low levels of the atmosphere may show important trends in storm-relative helicity. Also strengthening of upper-level flow may hint at upper-level dynamics and increasing bulk shear moving into the threat area.

The VWP may indicate coupling or decoupling of the boundary layer, especially at night or during the morning. This is most obvious when lower-tropospheric winds remain constant but winds in the lowest 1 to 2 thousand feet drop to calm after dark or increase to match the lower-tropospheric winds during the morning.

Also frontal inversions can be assessed using the VWP product, signified by two layers with markedly different wind regimes. The VWP can reveal the depth and character of the cold air mass. If the front itself is too close to the radar site, though, the radar may have trouble finding a representative wind figure in the lowest level.

a closer look

VWP shows a profile plot of winds aloft very much like wind profiler systems. However the sample is volumetric, covering a region nearly 200 miles wide, and so important features may be smoothed over. Another significant difference is the VWP requires many more scatterers than wind profilers.

VWP requires a significant number of scatterers to work. The best scatterers are produced by dust, insects, and cloud droplets. If scatterers are not present, a data void will occur. This is quite common, especially during good weather.

Bad data will be produced by birds, air mass boundaries, and thunderstorms. This will reflect velocity signatures that differ from the mean wind at that level. It may increase the RMS error at that level or cause it to be rejected altogether. Use the results with extreme caution when any of these are suspected.

The VWP product is not used as often as it should be for identifying trends in low-level shear that could increase the potential for severe thunderstorms.

A RMS error of over 9.7 kt or a symmetry error of over 13.6 kt will invalidate the data for that level.

For a given level, the highest amplitude of the VWP sine wave is considered to be the wind speed, while the phase of the highest inbound amplitude is considered to be the wind direction.

A void will occur when any data sample fails the VWP algorithm.

websites

weather.cod.edu/analysis/radar.main.html
www.nws.noaa.gov/radar
www.weathertap.com (pay)

VAD Wind Profile (VWP). This graph shows wind speed and direction as a function of altitude (Y) and time (X). In this example, the column on the far right represents the most current observation. Winds are southeasterly at about 15 kt near the surface and 20 to 30 kt in the upper troposphere. Note the color coding used for the RMS error. It is important to examine the scale, shown here at the bottom, to understand what colors are being used for good data.

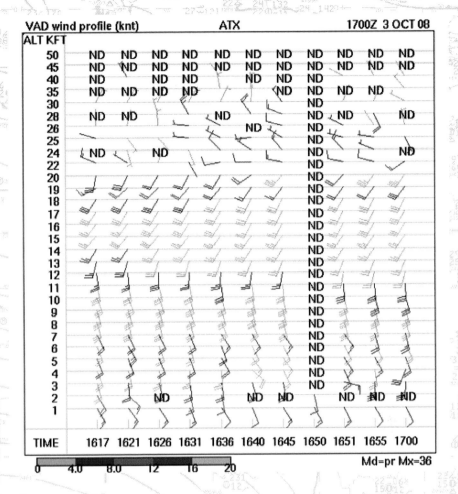

Velocity Azimuth Display (VAD). Not to be confused with a Velocity Wind Profile (VWP), this is what a velocity azimuth display really looks like, as obtained from a hardcopy at a WSR-88D site. It is a graph of azimuth (X) versus velocity (Y) at a given height, in this case 13,000 ft. This is a primitive radar product and is unlikely to be found on the Internet since a single volume scan could generate dozens of these graphs. However it serves a useful example by showing the sine wave and data points from which a VWP product is constructed.

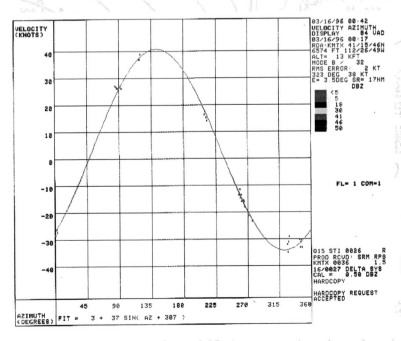

Free text message

The radar is down. Fine, but how do you find out what the cause was and when it is expected to be back up? What if there is some important change in the operating parameters you need to know about? With the 1970s-era legacy radar network, details about outages were handled by phone calls. The outside world only knew about problems via teletype RAREP reports indicating PPINA, PPIOM, or PPINO for that site. As broadcasters, military users, and flight service stations began tapping into the radar network via Kavouras dialup boxes in the late 1980s, these messages did little to answer questions about what was going on at a particular site.

Fortunately the WSR-88D designers had a plan in mind: the Free Text Message (FTM). This product is available for all radar sites, and it gives radar operators tools and procedures to post public information pertinent to the radar. This reaches the hands of not only hobbyists and media forecasters, but also air traffic control centers, other weather offices, and the NOAA Radar Operations Center. It also allows the operators to post expected stop and start times.

During the early and mid-1990s, processing hardware was quite expensive. When the radar product generator was stalling and loadshedding was occurring, the Free Text Message allowed operators to send out a status message. It would usually encourage users to cut back on product requests, which were queued at the radar site and could stack up tremendously during severe weather.

A useful link to check the status of the entire United States radar network <*weather.noaa.gov/monitor/radar*>. The page contains a list of all radars color-coded by latency of data. From there, users can view Free Text Messages at individual sites. The page automatically updates every 60 seconds.

```
. . . . . . . . . . . . . . . . . . . . . . . . . . . . . . . . . . . . . .
Mar 18 2006 05:50:30
NOUS64 KFWD 180550
FTMFWS

WE HAVE LOST THE WIDEBAND CONNECTION
WITH THE KFWS WSR-88D. TELCO IS BEING
NOTIFIED BUT WE HAVE NO ETA FOR RETURN
TO SERVICE. MOORE, FIC WFO FWD
. . . . . . . . . . . . . . . . . . . . . . . . . . . . . . . . . . . . . .
```

a closer look

Free Text Messages (also sometimes called Status Messages) reveal important information about outages, defective equipment, and other conditions that may affect the radar unit. It is sent over the standard NOAAPORT text feed.

When a radar is not generating any products or the data looks erroneous, look at the Free Text Message to find what is causing the problem and to determine when the site will be back online.

NWS Instruction 10-2201 designates the use of free text messages. It states: "Should the radar become nonoperational, the back up office will send a free text message notifying users of the outage and expected time of return to service. In this case, the backup office will use neighboring radars, including supplemental WSR-88Ds (Air Force), which have overlapping coverage to cover the shortfall."

Free Text Messages are aggregated at the NOAA Radar Operations Center (ROC) in Oklahoma, which operates a website showing the nationwide radar status and can help suggest alternative sites.

During much of the 1990s FTM messages could only be entered on a UCP (Unit Control Position) terminal. Many of the radar tasks have now shifted to the AWIPS workstation system, where messages are now typed.

If your Web radar service does not provide Free Text Messages, ask for them. They are an important part of the WSR-88D feed.

The product code is 75/FTM.

websites

weather.cod.edu/analysis/radar.main.html
http://www.weathermatrix.net/archive/FTM/
www.nws.noaa.gov/radar
www.weathertap.com (pay)

```
. . . . . . . . . . . . . . . . . . . . . . . . . . . . . . . . . . . .
000
NOUS64 KEWX 260612
FTMDFX

THE RADAR WILL BE DOWN UNTIL FURTHER NOTICE DUE TO AN UNEXPECTED
OUTAGE. SORRY FOR THE INCONVENIENCE. 26/7Z

. . . . . . . . . . . . . . . . . . . . . . . . . . . . . . . . .
000
NOUS64 KEWX 260843
FTMDFX

THE KDFX RADAR IS OUT OF SERVICE FROM A DIRECT HIT BY A SEVERE
THUNDERSTORM.   RADOME DAMAGE HAS OCCURRED.   IT IS UNKNOWN HOW LONG
THE RADAR WILL BE OUT OF SERVICE BUT IT WILL TAKE SEVERAL DAYS TO
ASSESS THE OVERALL DAMAGE AND MAKE REPAIRS.   345 AM CDT MAY 26 2001
JDW

. . . . . . . . . . . . . . . . . . . . . . . . . . . . . . . . . . .
000
NOUS64 KEWX 130723
FTMDFX

THE KDFX RADAR IS BACK IN SERVICE.     13 JULY 2001   PASHOS, FIC

. . . . . . . . . . . . . . . . . . . . . . . . . . . . . . . . .
```

On May 26, 2001 at 1:10 am CDT the Del Rio, Texas WSR-88D radar was hit with 80 mph wind and hail, destroying the 39-foot diameter rigid fiberglass radome. It took 49 days to bring the site back online. The Free Text Message product shows how users were notified of the outage and repairs. *(NWS photo)*

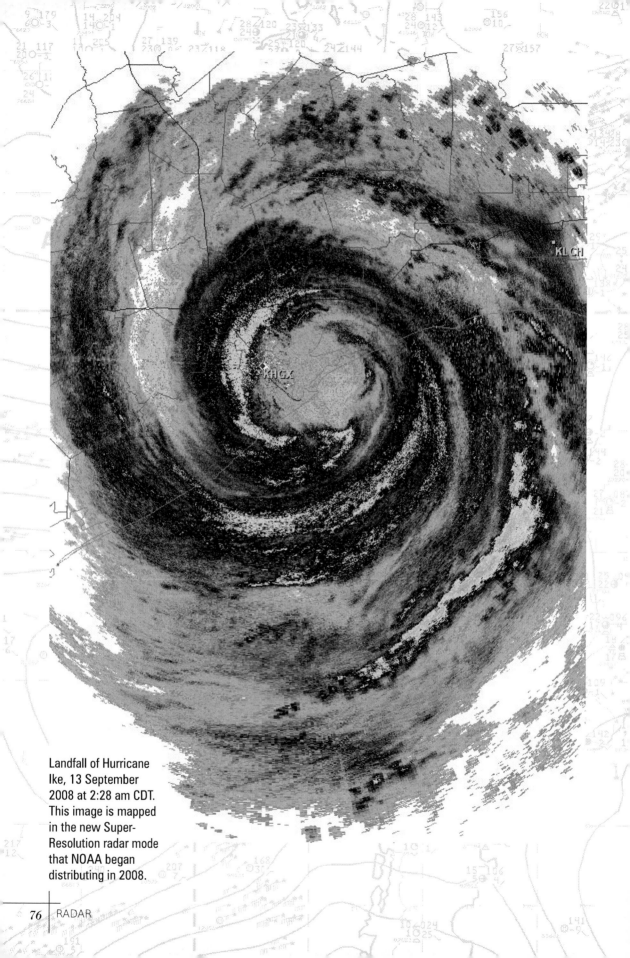

Landfall of Hurricane
Ike, 13 September
2008 at 2:28 am CDT.
This image is mapped
in the new Super-
Resolution radar mode
that NOAA began
distributing in 2008.

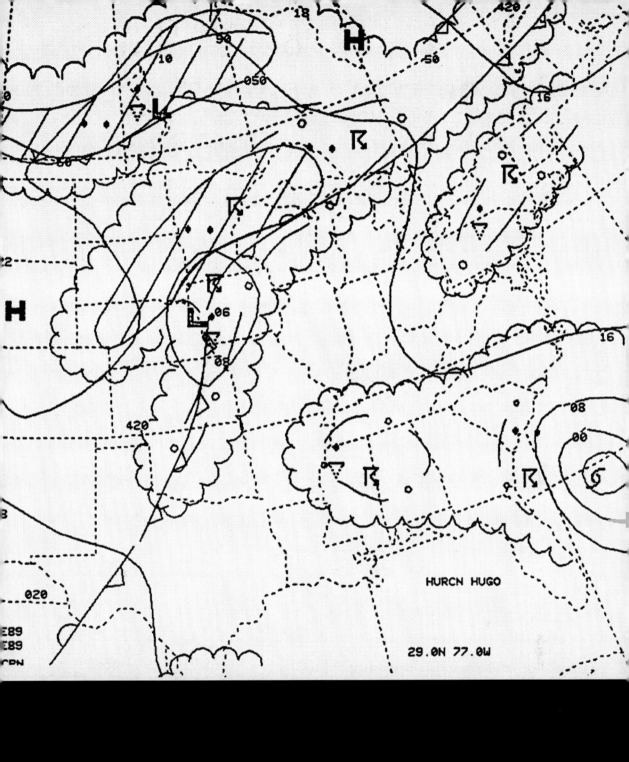

HURCN HUGO

29.0N 77.0W

SPC Convective Outlook

The SPC Convective Outlook is an official forecast transmitted four times a day which establishes the threat of severe thunderstorms across the United States. It's created by a qualified team of thunderstorm forecasters operating within a special branch of the National Oceanic and Atmospheric Administration: the Storm Prediction Center (SPC) in Norman, Oklahoma. The product was first transmitted by the Severe Local Storms unit (SELS) in 1955 and by fax in 1961 from Kansas City.

Currently, a vast array of observational data and models are used to develop the convective outlook. This is comprised of every forecasting tool available, including surface and upper air observations, profiler data, satellite data, and RUC, Eta, and MM5 model output.

> "PROGS CONT UNSTBL AMS ALG LEE SLOPES WITH UPSLOPE FLOW AT LO LVLS AND WK S/WV MVG ACRS MTNS...WK PVA SHUD TRIGGER UNSTBL AMS AHD OF SFC BNDRY LEFT BY NOCTURNAL TSTMS."
>
> - SELS, June 1979

Convective outlooks come not only with a map but also with an informative text bulletin. It defines the reasoning behind the outlook and provides a brief overview of what is expected to unfold. The outlook coordinates included in the bulletin allow human plotters and machines to duplicate the bulletin's coordinate information.

The Convective Outlook has been a favorite tool of storm chasers since the 1980s. During this era it appeared every morning on the PBS *A.M. Weather* program, suggesting the action area of the day.

Other products from SPC

SPC's products narrow down to two other types of products when the event draws closer. The Mesoscale Discussion (MCD) is disseminated when conditions appear to be favorable for severe thunderstorm development within the next three hours. It is usually transmitted when an imminent development area is identified, such as when an enhanced towering cumulus field begins developing in a suspected target area.

Finally, when a public advisory is required and severe weather is imminent, a severe thunderstorm or tornado watch is issued. It generally has a lifespan of four to six hours. Watch boxes considered by the SPC forecaster to be associated with an "enhanced risk of very severe and life-threatening weather" are transmitted as PDS watches, short for *particularly dangerous situation*. Only about 3% of all watches receive a PDS designation.

risk definitions

General. There is a 10-percent or better chance of thunderstorms.

Slight risk (SLGT). Well-organized severe thunderstorms are expected but in small numbers and/or coverage. Verification is made up of either 5 to 29 reports of 1-inch or larger hail, and/or 3 to 5 tornadoes, and/or 5 to 29 wind events.

Moderate risk (MDT). Great concentration of severe storms. Verification is made up of either 30 or more reports of 1-inch or larger hail, and/or 6 to 19 tornadoes, and/or 30 or more wind events.

High risk (HIGH). A major severe weather outbreak. At least 20 tornadoes are expected with at least two producing F3 damage or an extreme derecho event with widespread wind damage.

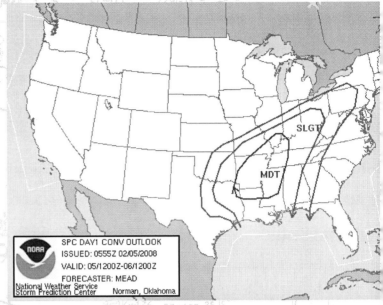

SPC DAY1 CONV OUTLOOK
ISSUED: 0555Z 02/05/2008
VALID: 05/1200Z-06/1200Z
FORECASTER: MEAD
National Weather Service
Storm Prediction Center   Norman, Oklahoma

Above right: A sample convective outlook preceding the tornado outbreak of February 5, 2008. Indeed the moderate risk is not something to be taken lightly.

Center right: Forecasting floor at the Storm Prediction Center, as seen during a graveyard shift in February 1997. *(Tim Vasquez)*

Below right: In this 1976 photo we see Don Wales, Tim Oster, Larry Wilson, and Bill Henry developing the afternoon forecast at the Severe Local Storms Unit in Kansas City. *(SPC archive)*

HPC products

The Hydrometeorological Prediction Center (HPC) in Camp Springs, Maryland assumes much of the centralized forecast guidance for United States weather offices. It came into being in late 1995 with the reorganization of the National Meteorological Center (NMC) into the National Centers for Environmental Prediction (NCEP). The main objectives of HPC are to provide quantitative precipitation forecasts, medium-range public forecasts, numerical model diagnostics and discussions, surface analysis responsibilities, and an international desk for visiting meteorologists.

The national forecast graphics are a useful chart for forecasters who want a quick summary of expected weather across the nation. They are often copied verbatim by broadcast media outlets to show the next day's weather. The graphics provide excellent information on the extent of precipitation areas and the expected transition lines between rain and snow.

The medium range forecast products, extending out to 7 days, are prepared by a team of two meteorologists twice a day. One forecaster focuses on weather systems and fronts while the other concentrates on precipitation and temperatures. A preliminary graphic is released to National Weather Service offices at 10 am EST for coordination purposes, with the public release occurring at 2 pm EST.

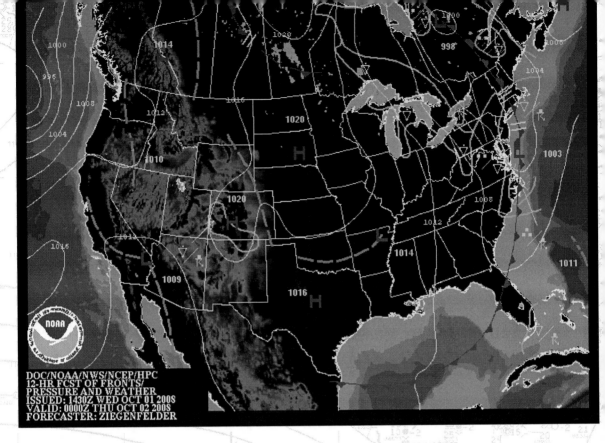

DOC/NOAA/NWS/NCEP/HPC
12-HR FCST OF FRONTS/
PRESSURE AND WEATHER
ISSUED: 1430Z WED OCT 01 2008
VALID: 0000Z THU OCT 02 2008
FORECASTER: ZIEGENFELDER

MODEL DIAGNOSTIC DISCUSSION / NWS HYDROMETEOROLOGICAL PREDICTION CENTER CAMP SPRINGS MD
125 PM EDT WED OCT 01 2008 / VALID OCT 01/1200 UTC THRU OCT 05/0000 UTC

MODEL INITIALIZATION...
ANY INITIALIZATION ERRORS WITH THE NAM AND GFS DO NOT APPEAR TO
SIGNIFICANTLY IMPACT THEIR FORECASTS.

MODEL TRENDS...
...SHORTWAVE TROUGH SWINGING THROUGH EASTERN U.S. DAY 1... COMPARED TO ITS 00Z RUN...ON DAY 2 THE NAM LIFTS A
SLIGHTLY MORE AMPLIFIED SHRTWV FURTHER TO THE E ALONG THE NORTHEAST COAST...ALLOWING MORE PHASING WITH THE
SYSTEM OVER THE GREAT LAKES AND IN TURN SHOWS THE DEVELOPMENT OF A STRONGER SFC WAVE LIFTING ACROSS NRN NEW
ENG AND A SECOND H5 LOW OVER ERN CANADA THAT WAS NOT FORECAST BY THE 00Z RUN. THE GFS ALSO LIFTS A SLIGHTLY
MORE AMPLIFIED SHRTWV NWD ALONG THE NORTHEAST COAST AND DEPICTS THE DEVELOPMENT OF A MORE DEFINED WAVE
DEVELOPING ALONG THE TRIPLE POINT.

Top: A human-created official forecast for 12 hours in the future.

Center: One of the many useful text discussions produced by HPC is the model diagnostic discussion, which outlines problem areas with the forecast models.

Below right: QPF precipitation forecast for the entire 120 hour period in the future.

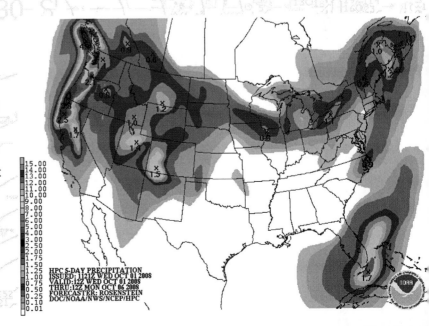

HPC 5-DAY PRECIPITATION
ISSUED: 1142Z WED OCT 01 2008
VALID: 12Z WED OCT 01 2008
THRU:12Z MON OCT 06 2008
FORECASTER: ROSENSTEIN
DOC/NOAA/NWS/NCEP/HPC

Area forecast discussion

There is probably no better insight into a National Weather Service forecaster's mind than the Area Forecast Discussion (AFD). These bulletins are composed before a forecast package is issued, typically twice a day with updates two more times per day.

The AFD was known as an SFD (State Forecast Discussion) before a restructuring initiative in the mid-1990s. It is also sometimes known as an FP, or FPUS3 bulletin, taking this name from its assigned header under the Family of Services (FOS) distribution circuit.

For decades the forecast discussion bulletin used a vast number of contractions. This characteristic carried over from the 1970s when data circuits were slow and brevity was valued. During the mid-1990s most offices began transitioning to plain English format. However, contractions are still frequently encountered, and they may be decoded with the guide shown to the right.

a closer look

The Area Forecast Discussion (AFD) is designed to allow the forecaster to explain their forecast thinking to other meteorologists, to weather-sensitive officials, and to interested members of the public. It is one tool that allows interoffice coordination of forecast thinking.

Who wrote the bulletin? The text is usually followed either by the forecaster's last name or an internal code identifying the forecaster.

A survey of 188 AFD users in 2001 by the Gaylord, Michigan WSFO showed that 69% were weather enthusiasts and 14% broadcasters.

websites

iwin.nws.noaa.gov

Below: Forecast discussions have been a part of NWS forecasting operations for decades. This example was about 20 years old at publication time. The use of contractions was much more common due to limited data bandwidth.

```
KFTW 160949
NORTH TEXAS FORECAST DISCUSSION
NATIONAL WEATHER SERVICE FORT WORTH TX
350 AM CST MON JAN 16 1989

MODELS IN GENL AGREEMENT IN KEEPING FAST ZONAL FLOW ACRS N AM WITH FLAT RDG
IN ERN PACIFIC AND FLAT TROF POSN ACRS CNTRL/E CNTRL U.S. WITH UPR STORM
TRACK SHIFTING NWD TO THE NRN PLAINS...AND AMPLIFICATION OF THE RDG/TROF
PATTERN TOWARDS THE END OF THE 48 HR FCST PD...LTL CHC FOR ANY SIG WX N TX
AREA. THIS MRNG'S ST DECK SHOULD CONT TO MOV S AND BRK UP THIS AFTN AS DRY
SUBSIDING AIR CONTS TO FEED INTO AREA FM THE NW.  SOME CI STREAKS WL CONT
TO MOV ACRS IN RELATIVELY STG UPR FLOW...BUT CHANNELED NATURE OF WEAK VORT
MAX'S WL INSURE LTL ELSE IN THE WAY OF SIG VERT MOTIONS/CLDS. AM A BIT CON-
CERNED ABT MSTR RETURN MOVG UP RIO GRANDE VALLEY AS ADVERTISED BY LFM AND SM
ON DAY 2...BUT MSTR WOULD HAVE TO TURN A TIGHT ANTICYCLONIC CORNER TO MAKE
IT INTO N TX...MORE THAN LIKELY ANY CLDS WOULD BREAK UP OR MOV EWD IN SUCH
A SITUATION. STILL... ZONAL 3-WAVE PATTERN ARND THE NRN HEMISPHERE BEARS
WATCHING AS SIG ADJUSTMENTS ARE PSBL LATER IN THE WEEK.
ABI 59/32/64   001
ACT 58/32/62   001
SPS 53/25/62   001
.NT...NONE
#26
```

discussion of katrina

88
FXUS64 KLIX 291006
AFDLIX

AREA FORECAST DISCUSSION
NATIONAL WEATHER SERVICE NEW ORLEANS LA
506 AM CDT MON AUG 29 2005

.DISCUSSION...
EXTREMELY DANGEROUS CATEGORY FOUR HURRICANE KATRINA WILL BE MAKING
INITIAL LANDFALL OVER LOWER PLAQUEMINES PARISH SHORTLY. THE
PROJECTED TRACK TAKES KATRINA NORTH TOWARDS THE LOWER PEARL RIVER
VALLEY NEAR THE LA/MS BORDER LATE THIS MORNING. HURRICANE WINDS
HAVE BEEN EXPANDING THE LAST COUPLE HOURS. IN ADDITION TO VERY
STRONG WINDS...A MASSIVE STORM SURGE IS OCCURRING ALONG WITH
EXTREMELY HEAVY RAINFALL AND FLASH FLOODING. WILL HAVE TO CUT THE
DISCUSSION SHORT TO GET PRODUCTS OUT BEFORE LOSING COMMUNICATIONS.
REFER TO THE OUR LATEST HURRICANE LOCAL STATEMENT...FLOOD
WATCH...AND INLAND HURRICANE/ TROPICAL STORM WARNING PRODUCTS.

NO OTHER SIGNIFICANT WEATHER SYSTEMS ARE EXPECTED TO FOLLOW
KATRINA...SO HOT AND DRIER WEATHER IS EXPECTED TUESDAY THROUGH
SUNDAY.

&&

FORECAST DISCUSSION CONTRACTIONS

ABNDT abundant
ABT about
ABV above
AC convective outlook
ACLT accelerate
ACPY accompany
ACTV active
ADJ adjacent
ADL additional
ADVCT advect
ADVN advance
ADVY advisory
ACFTG affecting
AFT after
AFTN afternoon
ALG along
ALQDS all quadrants
AMD amend
AMS air mass
AMT amount
ANAL analysis
AOA at or above
AOB at or below
APRNT apparent
ARPT airport
ATTM at this time
AWT awaiting
BCM become
BD blowing dust
BDRY boundary
BFR before
BLD build
BLO below
BYD beyond
CAA cold air advection
CCLDS clear of clouds

CDFNT cold front
CG cloud to ground
CHC chance
CI cirrus
CIG ceiling
CNTY county
COR correction
CPBL capable
CRNR corner
CSDRBL considerable
CST coast
DBL double
DCR decrease
DFNT definite
DISC discussion
DLA delay
DLT delete
DMG damage
DPND deepened
DSCNT descent
DSIPT dissipate
DURG during
DVLP develop
DVRG diverge
DVRGG diverging
DVV downward motion
EBND eastbound
ELNGTD elongated
EMBDD embedded
ERN eastern
ERY easterly
ETA Eta model
EWD eastward
FA aviation forecast
FLRY flurry
FLW follow

FQT frequent
FRM form
FROPA frontal passage
FRZN frozen
FT terminal forecast
FWD forward
GEN general
GND ground
GRDL gradual
GRT great
HAZ hazard
HGT height
HLF half
HLSTO hailstones
HV have
HVY heavy
HWY highway
IC ice
IMDT immediate
IMPL impulse
INCL include
INCR increase
INDC indicate
INDEF indefinitely
INSTBY instability
INTS intense
IOVC in overcast
INVOF in vicinityof
IPV improve
ISOL isolate
ISOLD isolated
KFRST killing frost
KT knots
LAT latitude
LCL local
LCTMP little temp chg

LFTG lifting
LGRNG long range
LGWV long wave
LI lifted index
LIS lifted indices
LK lake
LKLY likely
LLJ low level jet
LMTD limited
LN line
LRG large
LST local standard time
LTD limited
LTG lightning
LTL little
LYR layer
MAX maximum
MDFY modify
MDL model
MDT moderate
MED medium
MEGG merging
MESO mesoscale
MET meteorological
MID middle
MISG missing
MOV move
MRGL marginal
MRNG morning
MSG message
MSL mean sea level
MST most
MXD mixed
NAB not above
NBND northbound
NEG negative

NGM nested grid model
NGT night
NIL none
NLY northerly
NNNN end of message
NRLY nearly
NTFY notify
NVA negative vorticity
advection
NXT next
OBND outbound
OBS observation
OBSC obscure
OCLD occlude
OCNL occasional
OCR occur
OFC office
OFP occluded frontal
passage
OMTNS over mountains
OTLK outlook
OTP on top
OTR other
OTRW otherwise
OVC overcast
OVTK overtake
PBL boundary layer
PCPN precipitation
PLNS plains
POS positive
PRST persist
PSBL possible
PTLY partly
PVA positive vorticity
advection
PVL prevail

QPF quantitative
precipitation forecast
RCV receive
RGL regional model
RGN region
RNFL rainfall
ROT rotate
RPLC replace
RSG rising
SA surface observation
SCT scatter/scattered
SEPN separation
SGFNT significant
SKC sky clear
SLP slope or pressure
SPRD spread
SR sunrise
SS sunset
STFR stratus fractus
SX stability index
TNTV tentative
TROF trough
TS thunderstorm
UPR upper
UVV upward motion
VC vicinity
VFR visual flight rules
VRG veering
VV vertical velocity
WAA warm advection
WKN weaken
WTR water
WX weather
YDA yesterday
ZN zone

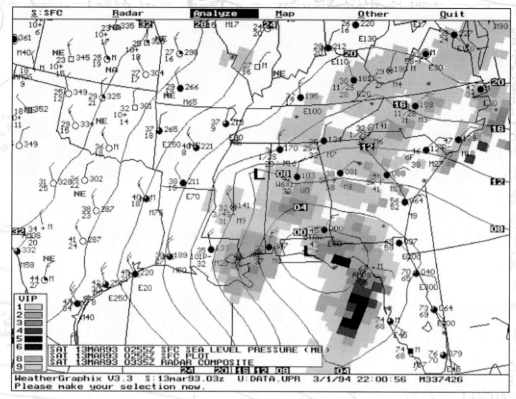

In the days before the Internet as we know it, hobbyists only had a limited assortment of data to work with. The graphis show surface data and radar on 13 March 1993, date of the so-called Storm of the Century.

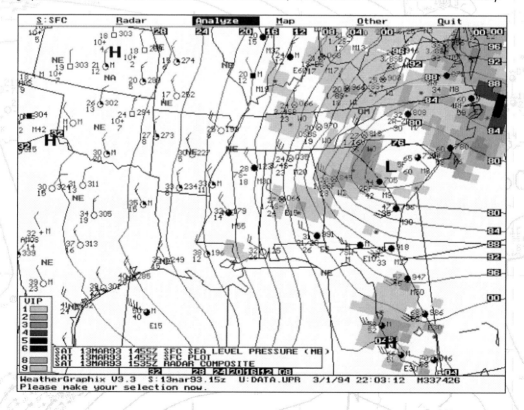

```
C - CALCULATE QDOT
C - SIGMADOT=-RHO.G.W/PS-SIG/PS.DPSDX.U-SIG/PS.DPSDY.V
C
      DO K=2,KL
        DO I=1,ILX
          RHOOS=TWT(K,1)*RHOO(I,J,K)+TWT(K,2)*RHOO(I,J,K-1)
          QDOT(I,J,K)=-RHOOS*G*W3D(I,J,K)*RPSA(I,J)*0.001
     +                   -SIGMA(K)*(DPSDXM(I,J)*(TWT(K,1)*UCC(I,J,K)
     +                                    +TWT(K,2)*UCC(I,J,K-1))
     +                        +DPSDYM(I,J)*(TWT(K,1)*VCC(I,J,K)
     +                                    +TWT(K,2)*VCC(I,J,K-1)))
        ENDDO
      ENDDO
C
C - CALCULATE TOTAL DIVERGENCE AND STORE IN DIVX
C
      DO I=1,ILX
        DUMMY(I)=1.0/(DX2*MSFX(I,J)*MSFX(I,J))
      ENDDO
C
      DO K=1,KL
        DO I=1,ILX
          DIV=UA(I+1,J+1,K)+UA(I,J+1,K)-UA(I+1,J,K)-UA(I,J,K)+
     +        VA(I+1,J+1,K)+VA(I+1,J,K)-VA(I,J+1,K)-VA(I,J,K)
          DIVX(I,J,K)=DIV*DUMMY(I)+(QDOT(I,J,K+1)-QDOT(I,J,K))*
     +                   PSA(I,J)/DSIGMA(K)
        ENDDO
      ENDDO
    ENDDO
C
C *** COMPUTE HORIZONTAL ADVECTION TERMS FOR U, V, PP, W:
C
      CALL HADV(KZZ,U3DTEN,UA,VA,U3D,MSFD,DX16,3,INEST)
      CALL HADV(KZZ,V3DTEN,UA,VA,V3D,MSFD,DX16,3,INEST)
      CALL HADV(KZZ,PP3DTEN,UA,VA,PP3D,MSFX,DX4,1,INEST)
```

CHAPTER 5

NUMERICAL PREDICTION

Numerical Weather Prediction

Numerical weather prediction is the science of predicting the future state of the atmosphere using physical equations. The technique was first proposed in 1923 by Lewis Frye Richardson, a British meteorologist, in his book Weather Prediction by Numerical Process. Unfortunately, the technology and resources were simply not available at the time to pursue the idea.

With the breakthroughs in computing technology during the 1950s, researchers had fantastic opportunities for laying down the craft of numerical weather prediction. Universities and government institutions worked together to develop simple barotropic models, which became more sophisticated as scientific knowledge grew. The first operational predictions were available by 1960, and were followed by fantastic leaps and bounds in the decades ahead.

What is a numerical model?

While the physics and dynamics of models are far beyond the scope of this book, it is most important to point out that each model runs at a unique scale. Depending on the weather agency and purpose, a model is either run at a coarse resolution covering the entire world, or is run at a fine resolution covering a small area (usually the agency's home continent). The worldwide model is called a "global" model, while the fine model is called a "regional" model. Even finer models covering smaller areas, such as portions of a country, are called "mesoscale" models.

Detailed models are very accurate but are limited by their enormous processing requirements, which limits them to a specific area of the globe. As a result they suffer from boundary errors on their edges and their forecasts begin degrading just one or two days into the future. During the first day or two, forecasters are encouraged to use the regional models for their greater detail, but past this point should use global models, which lack boundary errors.

Forecasters must also be aware of the unique characteristics of each model, which are detailed section by section in this book, where applicable. These can range from eccentricities in the parameterization of radiation and convection, which show up in unusual ways, to shortfalls in vertical and horizontal resolution which give the model weaknesses in certain weather regimes. All of these are referred to as model biases.

When the hemispheric wave number, the number of long waves around a hemisphere, undergoes a change, it is referred to as a wave number transition. A common example is when a low wave number suddenly increases from 2 or 3 to 5 or 6 over one to two days. One signal of a possible pending transition is when a very strong, long, broad polar jet becomes established in the North Pacific Ocean. Transition events signal a major shift in weather regimes, and are notorious for causing errors and inconsistencies in numerical weather forecasts. Always be wary of forecasts beyond 48 hours when a wave number transition is suspected or is underway.

Getting model data

There are four primary sources of model data graphics on the Internet: government, academic, commercial, and hobbyist. The exception is "link sites", which contain no original content and simply refer to one of the sources below. Though they may be handy for navigating to the right product, such sites are not included in this book as they are often incomplete, outdated, or abandoned. One exception is <www.westwind. ch> which has a surprisingly extensive set of links to international model runs not covered in this book.

Government agencies. Common sense says that getting it from the horse's mouth is the best thing. Quite often, however, the products placed online at various weather agencies for public consumption are not as pleasing, detailed, or

NCEP Computing Power

Year	Model	Proc	Speed (Mflops)	Memory (MB)	Disk storage	Operational Model
1956	IBM 701	1	0.001	0.001	9 KB	(Research)
1958	IBM 704	1	0.008	0.02	144 KB	Barotropic
1960	IBM 7090	1	0.067	0.1	50 MB	3-lvl QG model
1963	IBM 7094	1	0.1	0.1	50 MB	Baroclinic
1966	CDC 6600	1	3	1	75 MB	6-lvl PE, LFM
1974	IBM 360/195	1	10	4	300 MB	LFM, 7-lvl PE
1978	CDC Cyber 205	1	100	32	7.2 GB	OI, GSM, NGM
1987	CDC Cyber 205	2	200	64	14.4 GB	OI, GSM, NGM
1990	Cray Y/MP8 VII	1	2,600	512	2 GB	NGM, ETA
1994	Cray C90/16256 [1]	16	15,300	2 GB	200 GB	NGM, ETA, RUC
1999	IBM RS/6000 SP	768	700,000	192 GB	4.6 TB	ETA, RUC2, etc
2000	IBM RS/6000 SP	2048	2,500,000	256 GB	7.5 TB	ETA, RUC2, etc
2003	IBM pSeries 690	44	7,300,000	1.4 TB	42 TB	ETA, RUC2, etc
2004	IBM pSeries 655	1152	12,456,000	6 TB	91 TB	GFS, NAM, etc
2006	IBM p575 [2]	2496	15,470,000	5 TB	160 TB	GFS, NAM, etc

HOME COMPUTER COMPARISION

Year	Model	Proc	Speed (Mflops)	Memory (MB)	Disk storage	Operational Model
1981	Commodore 64	1	0.16	0.064	144 KB	—
2008	Intel Core 2 Quad	4	1200	3 GB	500 GB	Desktop ETA

[1] Caught fire 27 Sep 1999 and was destroyed. Models ran on backup Cray C5 until replaced by IBM RS/6000 SP on 18 Nov 1999.

[2] Upgraded to 54 million megaflops by 2008.

Some information obtained from "Maturity of Operational Numerical Weather Prediction: Medium Range", 1998, by Eugenia Kalnay, Stephen J. Lord, and Ronald D. McPherson, Bulletin of the American Meteorological Society.

functional as those on other sources.

Universities and institutions. Academic sources are often eager to demonstrate that they are on the cutting edge, and there you can find some of the best products available anywhere. The College of DuPage <*weather.cod.edu*> and UCAR <www.rap.ucar.edu> are well-known examples of sites containing excellent product lines. On the other hand there are more than a few universities with excellent meteorology programs but with a neglected or non-existent weather server.

Commercial sources. For those willing to pay a reasonable fee, a handful of companies generate products in-house from original model output. One of those is Wright Weather <*www.wright-weather.com*>, which offers some unusual guidance for the United States such as the Japanese JMA/GSM and UKMET model.

Hobbyist. With GRIB software and some models available, some hobbyists and noncommercial enthusiasts have begun putting model charts on the Internet. One popular source is Earl Barker's site at <*www.wxcaster.com*> which has one of the largest collections of model graphics available for North America.

The vast majority of Internet weather graphics are generated using GEMPAK, Unisys WXP, or GRADS software. Because of this, it's quite common to notice a familiar look that seems to be shared between two very different sites. For example, NIU <*weather.admin.niu.edu*> and Unisys <*weather.unisys.com*> have products with a similar look since they both use the Unisys WXP engine. Government weather sites tend to have complex display software developed inhouse or on contract, and may have very unique and surprising appearances, for better or worse.

Other global models

Finally it should be mentioned that this is not the complete inventory of global models which are available. Dozens of nations have operational global models, including Germany, Russia, China, and India. However they are generally not included in this book either because of translation barriers or because the host agency will not share their graphics publically. Furthermore, for model data that does exist publically, most non-American models have neither detailed fields nor gridded output data available for further inspection.

During the past few decades, the United States has embraced a truly free interchange of model data and modelling software, a policy which accelerated around 2000 with the modernization of the government Internet infrastructure. The implications of this cannot be understated. As 2010 approaches, this philosophy is paying massive dividends for hobbyists and the private sector.

A forecast of isentropic surfaces for Sweden, for example, has never been available from the ECMWF or UKMET model to the general public, as European weather agencies consider it restricted data. But real-time data from the ECMWF's counterpart in America, the GFS, is freely available in its entirety on the Internet. With the right website or the right software, those isentropic forecasts of Sweden can be made with American GFS data. Furthermore, thanks to a NCAR/NOAA partnership which created the WRF and placed it in the public domain, any hobbyist with a Linux machine can run a cutting-edge real-time mesoscale model just for Sweden and get not just isentropic fields but high-resolution model output with significant meteorological value.

That said, it is important to remember that numerical models, while often providing useful forecast solutions, tend to contain varying amounts of errors and biases and cannot be used to develop a sound, dependable forecast without an actual diagnosis of the atmosphere by a skilled forecaster. Using a model by itself is like predicting a close-race election based on polls. By itself, the result is just an uninformed guess without context, understanding, and insight.

Right: 3-dimensional displays of meteorological fields are sometimes encountered, but the technique is rarely used among operational meteorologists. Geographical characteristics are disorienting, data is often obscured, and it is difficult for the brain to assimilate the information. Caveat emptor!

an excerpt from

Weather Prediction by Numerical Process

L. F. Richardson, 1922

It took me the best part of six weeks to draw up the computing forms and to work out the new distribution in two vertical columns for the first time. My office was a heap of hay in a cold rest billet. With practice the work of an average computer might go perhaps ten times faster. If the time-step were 3 hours, then 32 individuals could just compute two points so as to keep pace with the weather, if we allow nothing for the very great gain in speed which is invariably noticed when a complicated operation is divided up into simpler parts, upon which individuals specialize. If the co-ordinate chequer were 200 km square in plan, there would be 3200 columns on the complete map of the globe. In the tropics the weather is often foreknown, so that we may say 2000 active columns. So that 32 x 2000 = 64,000 computers would be needed to race the weather for the whole globe. That is a staggering figure. Perhaps in some years' time it may be possible to report a simplification of the process. But in any case, the organization indicated is a central forecast-factory for the whole globe, or for portions extending to boundaries where the weather is steady, with individual computers specializing on the separate equations. Let us hope for their sakes that they are moved on from time to time to new operations.

After so much hard reasoning, may one play with a fantasy? Imagine a large hall like a theatre, except that the circles and galleries go right round through the space usually occupied by the stage. The walls of this chamber are painted to form a map of the globe. The ceiling represents the north polar regions, England is in the gallery, the tropics in the upper circle, Australia on the dress circle and the antarctic in the pit. A myriad computers are at work upon the weather of the part of the map where each sits, but each computer attends only to one equation or part of an equation. The work of each region is coordinated by an official of higher rank. Numerous little "night signs" display the instantaneous values so that neighbouring computers can read them. Each number is thus displayed in three adjacent zones so as to maintain communication to the North and South on the map. From the floor of the pit a tall pillar rises to half the height of the hall. It carries a large pulpit on its top. In this sits the man in charge of the whole theatre; he is surrounded by several assistants and messengers. One of his duties is to maintain a uniform speed of progress in all parts of the globe. In this respect he is like the conductor of an orchestra in which the instruments are slide-rules and calculating machines. But instead of waving a baton he turns a beam of rosy light upon any region that is running ahead of the rest, and a beam of blue light upon those who are behindhand.

Four senior clerks in the central pulpit are collecting the future weather as fast as it is being computed, and despatching it by pneumatic carrier to a quiet room. There it will be coded and telephoned to the radio transmitting station.

Messengers carry piles of used computing forms down to a storehouse in the cellar.

In a neighbouring building there is a research department, where they invent improvements. But these is much experimenting on a small scale before any change is made in the complex routine of the computing theatre. In a basement an enthusiast is observing eddies in the liquid lining of a huge spinning bowl, but so far the arithmetic proves the better way. In another building are all the usual financial, correspondence and administrative offices. Outside are playing fields, houses, mountains and lakes, for it was thought that those who compute the weather should breathe of it freely.

Ensemble predictions

Ensemble predictions are collections of two or more numerical weather prediction solutions, valid for the same location and forecast time. Combined with each other, each solution, called a member, gives a range of possible outcomes.

Given a set of starting conditions, a forecast model will always come to the exact same conclusion, even if it is run a century into the future. However to have a range of outcomes, it is necessary to introduce variability into the model. How can this be done?

The answer is to assume there are slight errors in the starting analysis, and make subtle adjustments before each run. This is called a perturbation for initial conditions (sometimes abbreviated "IC"). However, another approach is to assume that the model's approximations in its equations or algorithms are a greater cause of errors. In this case, the analysis field is held constant and the model is run multiple times with different physics, dynamics, and parameterizations. This is known as perturbation for the model's physics. In many cases ensembles are created with a blend of each type of perturbation.

Surprisingly, ensemble forecasting has been done for well over 35 years, in the tried-and-true method where forecasters compare the LFM to the NGM, the RUC to the Eta, and the Spectral to the ECMWF run. This is sometimes referred to as the "poor man's ensemble". However, many experienced forecasters still believe there is enormous value in comparing different models, and have expressed a desire for different model types and their ensembles to be included as part of an ensemble package.

Ensembles can be displayed in ways such as:

Ensemble mean. This chart is a mathematical average of all available model solutions for a given point in time. The results tend to look excessively smoothed, however the consensus on weather feature locations are immediately apparent.

Spaghetti diagram. This type of diagram shows one or two selected isopleths from each model combined together on one map. The isopleths look like long noodles of spaghetti overlapping one another, thus the name "spaghetti diagram". As a general rule, the more spread out the isopleths are, the weaker the confidence in the forecast for that area.

Standard deviation. A measure of the difference in all of the fields can be computed to show standard deviation. Where values are higher, there is less confidence in that area.

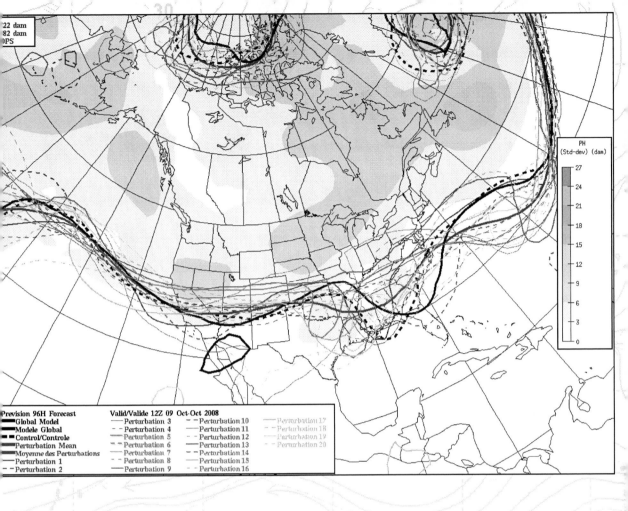

Prevision 96H Forecast   Valid/Valide 12Z 09 Oct-Oct 2008

Global Model	Perturbation 3	Perturbation 10
Modele Global	Perturbation 4	Perturbation 11
Control/Controle	Perturbation 5	Perturbation 12
Perturbation Mean	Perturbation 6	Perturbation 13
Moyenne des Perturbations	Perturbation 7	Perturbation 14
Perturbation 1	Perturbation 8	Perturbation 15
Perturbation 2	Perturbation 9	Perturbation 16

Perturbation 17
Perturbation 18
Perturbation 19
Perturbation 20

PH
(Std-dev) (dam)

Top: Spaghetti ensemble from CMC GEM runs. *(CMC)*

Right: NCEP SREF ensemble for sea level pressure, with mean pressure (isopleths) and standard deviation (color). The NCEP ensemble site allows flexible viewing of a number of ensemble parameters. *(NCEP)*

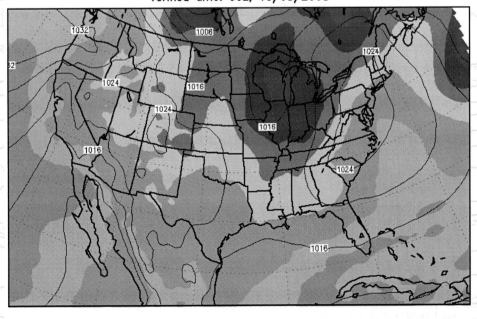

COM_US SLP(MB) 69H fcst from 09Z 05 oct 2008
mean is in contour and color represents spread in mb
verified time: 06z, 10/08/2008

0.1   0.2   0.5   1   1.5   2.5   3.5   4.5   6   8

Model soundings

Since models compute expected weather parameters at a large number of vertical points in the atmosphere, it is logical that a thermodynamic diagram of the data can be constructed. This was no easy task in the 1980s, even for National Weather Service field offices, and even in the early 1990s it was difficult to construct model forecast soundings without using the FDUS winds aloft products which were designed for pilots. However, nowadays the rapid advances in computing technology have made model output in sounding format a rather easy task.

There are a multitude of possible uses of model forecast soundings. For example, the sounding depiction format allows cold and warm layers to be properly evaluated, giving a detailed picture of winter precipitation type. In severe weather situations, the depth and relative strength of the elevated mixed layer (EML, the cap) is apparent. Heights of inversions, instability and shear parameters, and much more can easily be determined.

Model forecast soundings, however, are only as good as the model that produces them. A model with poor vertical resolution will produce poor vertical detail. Even the Eta has only 50 mb resolution, which provides conditions only about every 1500 ft (500 m) in the low levels. This can produce soundings that look excessively "smoothed" in convective weather situations and can smear out important, shallow layers. Also when strong gradients are present, such as when a sharp cold front is near a station, interpolations can produce excessively smoothed or unrepresentative results.

Furthermore, models do have difficulty with accurately forecasting the boundary layer due to the complex radiative, evaporative, and turbulent processes that occur there. These shortfalls are embodied in model forecast soundings.

Many National Weather Service offices have taken the initiative to identify and publish local studies relating model forecast sounding accuracy to actual soundings, in an attempt to benchmark the model sounding performance. The results are fairly promising but mixed. For example, one study (Evinson and Strobin, 1998) showed that the Eta forecast soundings were too moist in the low-levels, which caused overestimates of the instability in the Great Basin region of the United States.

Forecasters who rely on forecast soundings must be aware of their shortfalls, and must be alert to situations where the model, and in turn the sounding, is likely to fail.

a closer look

Model forecast soundings are an extraction of temperature and dewpoint data in a vertical column above a given point, within a given model run.

Model forecast soundings can be extracted from any numerical weather prediction model: Eta, GFS, ECMWF, MM5, and so forth. The only limitation is in the display software used to view the output, and the availability of the raw gridded data.

In a convective situation, the vertical resolution of a model forecast sounding is not sufficient to make precise forecasts about cap strength. Many models also have difficulty achieving this level of accuracy, where the difference of half a degree at 700 mb can make or break a severe weather event.

These soundings are highly sensitive to conversions from native grid resolution to degraded horizontal or vertical grids, such as when a field is remapped to 40 km or 80 km resolution.

The model forecast sounding is most prone to errors in the boundary layer, where there are a greater number of factors that can affect the sounding. These include radiation processes, topographical and vegetation effects.

Model soundings are heavily influenced by the convective precipitation scheme used within the model. Moisture and temperature effects from these schemes will advect to other areas and influence soundings at that location.

websites

www.stormchaser.niu.edu/machine/
  fcstsound.html
weather.cod.edu/fsound
www.arl.noaa.gov/ready/cmet.html
www.emc.ncep.noaa.gov/mmb/soundings.nam/
  snding.html
www.frd.fsl.noaa.gov/mab/soundings/java/
www.wright-weather.com (pay)

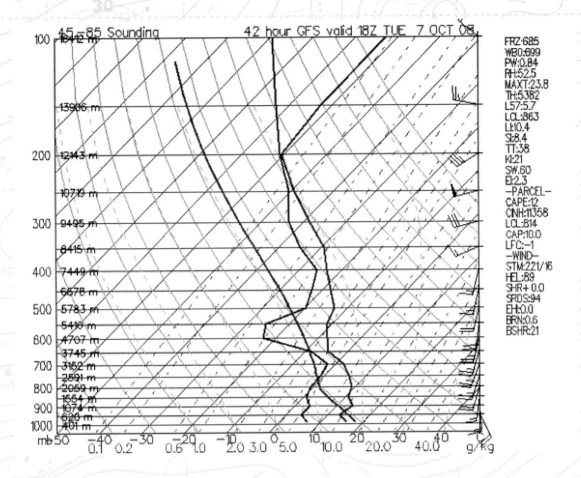

Above: GFS sounding at 45N 85W, in northern Michigan, from the College of DuPage site. Their site allows the ability to select from thousands of geographical points from a choice of any of NCEPs production models. *(College of DuPage)*

Right: NAM forecast sounding from NCEP. The NCEP site gives the ability to animate the forecast period and view temporal changes every six hours. *(NCEP)*

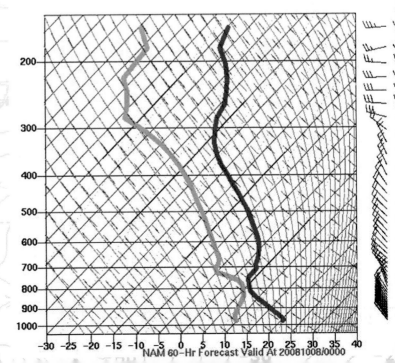

LFM model

The Limited-area Fine Mesh (LFM) model is completely gone from operational forecasting, but it left a deep inpact on North American forecasting from 1971 to 1993. It was the first regional (continental-sized) model to enter operational use.

Before the LFM, the National Meteorological Center (NMC, now NCEP) had relied on the 6LPE, or 6-layer primitive equation model, for most of its forecasting. This was a hemispheric model with a very coarse 381 km resolution. During the mid and late 1960s, the Air Force and NMC began work on developing a fine model that was windowed on continental areas where more data existed. The result was the LFM.

The LFM made its first operational forecast on 29 September 1971. It was initially a 6-layer model, with a resolution of 190.5 km (contrast this with 50 layers and 13 km with the current RUC!). The model was initially run only to 24 hours, but by 1975 and 1976 it was extended out to 36 then 48 hours. Computation on NMC's IBM 360 systems took about 20 minutes for the full run.

In 1979, a major upgrade to the LFM resulted in the LFM-II, which featured 7 layers and 127 km resolution, however in 1981 the earlier 190.5-km resolution was restored but added in a fourth-order finite difference approximation to compensate for the downgrade.

During the reign of the LFM in the 1970s and 1980s, it served as the basis for NMC to develop numerous sets of model output statistics (MOS), which allowed computers to automatically produce accurate forecasts for specific cities, taking into account LFM forecast fields, predictors, and empirical rules. The results were disseminated in the form of tables every 12 hours over teletype circuits.

By 1985, the end was in sight for the LFM as the cutting-edge NGM came online with almost twice the resolution in each dimension and much better accuracy. However the LFM excelled in its simplicity, so NMC assigned it a slot called the ERL (early) run, giving forecasters preliminary maps and charts just hours after radiosonde release time until the NGM package was available. The Cyber 205 systems took only 75 seconds to produce the full 48-hour LFM run.

The final nail in the coffin for the LFM came several years later as an even more sophisticated model, the Eta, made its introduction. The LFM was finally removed from the production suite in June 1993, replaced by the brand new Eta, and was completely phased out of all NCEP activities on 29 February 1996.

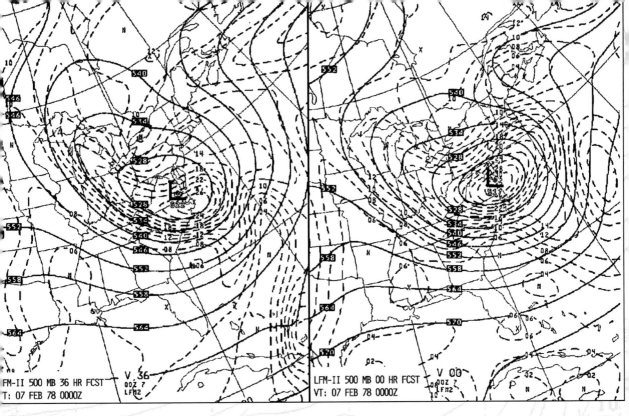

op: LFM analysis for 6 Feb 1978 at 0000 UTC from NMC showing a 36-hour forecast (left) and actual analyzed condi-
ons (right) of height and vorticity at 500 mb. Though the model had a good handle on the position of the vortex in
irginia, variations in the vorticity field can be seen. Note the difference in amplitude and position of the ridge over
exas and the trough over Ontario. In spite of these slight errors, this particular forecast of a 1978 snowstorm in the
ortheast United States was handled exceptionally well by the LFM.

elow left: Diagram showing the configuration of layers in the LFM (ERL) model during the mid-1980s, along with its
ibling models the spectral (MRF) and NGM (RAFS) model. *(NMC)*

elow right: Domain of the LFM model showing gridpoints and terrain The topography is significantly coarser than
aat used in modern forecast models. *(NMC)*

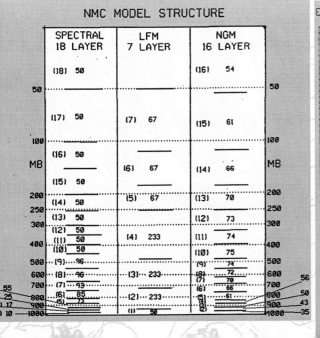

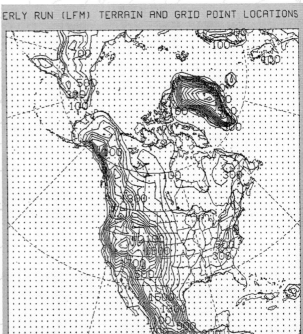

NGM model

The NGM (Nested Grid Model) was developed as the "next generation" improvement to the old LFM (Limited-area Fine Mesh model) that predominated United States forecasting in the 1970's. It was developed by Norman A. Phillips in 1978.

Operational use of the NGM commenced on 27 March 1985 on the new Cyber 205 supercomputer. An improved physics package was added in August 1986, followed by a fine-mesh grid expansion in February 1987. The model was scaled down to two grids in August 1991, and that concluded work on it with the bigger and better Eta expected to come online. The model was scheduled to be axed at NCEP by 1998, however this date has shifted to October 2009.

As implied by its name, the NGM uses multiple nested grids: a larger, coarse one to "see" distant systems in Asia, Europe, and elsewhere, and a smaller, denser one focused on North America in which to make highly detailed computations. It is a concept embraced by the Canadian GEM Regional and UKMET models. The NGM was part of a suite called the RAFS (Regional Analysis and Forecasting System), and is still sometimes referred to by that name.

Model biases

The NGM's dynamics have not been changed since 1990, and as a result a vast storehouse of NGM characteristics, both formal and anecdotal, have accumulated over the years. The most important biases are:

The skill of the NGM is best in the warm season when the atmosphere is less baroclinic.

Surface lows and highs are often too strong over land and too weak over the ocean. This is particularly true in the High Plains of North America.

The model has a poleward bias; in other words, United States systems often appear further north than they should. Systems emerging from the Rocky Mountains are often too far north and too intense.

Digging upper-level troughs show too weak of an amplitude.

Plunging cold outbreaks are moved too slowly and are preserved for too long.

The NGM forecasts spurious precipitation in the wake of squall lines in the southern Plains, and overdoes precipitation in upslope situations on the Great Plains and along sea breeze convergence zones. However it under-forecasts heavy rain events along the Gulf of Mexico coast.

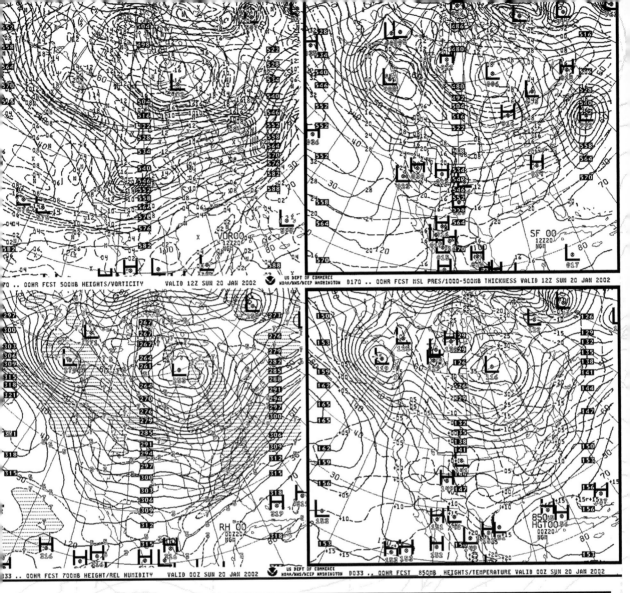

Top: In the 1980s and much of the 1990s, hobbyists and the private sector received NGM maps over the DIFAX facsimile system, typically in the 4-panel format seen above. The DIFAX format was not phased out until after 2000, and "old school" NGM charts were being produced as late as 2002 as shown here, with virtually no difference in their 1985 appearance.

Right: Domain of the current two-grid NGM, showing the large 336 km resolution outer grid (large box) and the smaller 84 km grid truncated to fit the Eta analysis grid (small, dark box). This configuration took effect in 1999. *(NCEP)*

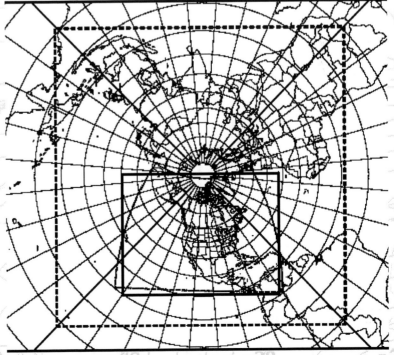

Eta model

The Eta model formed the backbone of most operational forecasting in the United States beginning in the mid-1990s. It was discontinued as a key NCEP model in June 2006 after the WRF/NMM was brought online to drive the NAM product suite. However it is still an important part of forecasting history and it is still being used as an ensemble member at NCEP, and its fields can still be viewed online. It is expected to be removed at NCEP by October 2009.

It is named after the greek letter "eta" and is pronounced EH-ta (not as the three alphabetical letters). The model was developed in 1978 at the University of Belgrade in Yugoslavia by Zavisa Janjic and Fedor Mesinger, using sigma surfaces to help negotiate the rough terrain in their home country. Janjic brought the model to NMC in the mid-1980s, and its success led to its official implementation in June 1993. The model was initially run at 00Z and 12Z with a resolution of 80 km at 38 layers. Output fields were generated out to 48 hours. The resolution was boosted to 48 km on 12 October 1995 with major physics improvements.

The NCEP model resolution is 12 km. From this solution, output fields on grids measuring 20, 40, and 80 km in resolution are generated. Thus the term "80 km Eta run" may apply to an output grid but does not correctly describe the model as it is currently run today. In other words, you are receiving the full benefit of the 12 km prediction scheme, but are looking at a coarser grid.

Model biases

The Eta model has proven to be an outstanding model. It is more accurate than the NGM, and less prone to convective feedback ("model blowup") compared to the GFS and older NGM. However it comes with a few biases that have been documented.

During the 1990s, the Eta carried a drying bias during the late spring where it dried out the air mass during the day on the eastern Great Plains, shunting the dryline too far east. Starting around 1998 the Norman OK weather office noticed a moist bias, a reversal of the old problem.

The Eta has shown a tendency to flip-flop on solutions beyond the 48-hour period. A prudent approach is to rely on the GFS for medium-range periods.

The Eta model, because of its use of eta coordinates, may not have enough vertical resolution over elevated terrain to process the boundary layer conditions correctly.

a closer look

The Eta model was removed from the NCEP forecast suite in 2006. It is still one of the members of the SREF (short range ensemble forecast) system but will likely be phased out completely in 2009.

The Eta model was considered to be cutting edge technology in the 1990s, at a time when forecasters only had access to the LFM, NGM, and the coarser MRF.

The Eta model is a finite difference (gridpoint) model that features 12 km resolution. It uses 60 pure eta levels, which are nearly horizontal but mathematically do not intersect the ground.

In a series of experiments in the 1990's, when the Eta model was test-run with sigma coordinates, currently used by the NGM and GFS models, it produced an erroneous poleward bias to frontal systems emerging from the Rockies, too-slow polar outbreaks, and cutoff lows too far east in the southwest U.S.

Eta-coordinate models are predisposed to underestimating boundary-layer winds in high terrain, and may miss katabatic windstorm situations.

CAPE values in the NCEP Eta model are found by lifting the parcel with the highest theta-e in the lowest 70 mb of the atmosphere.

Run codes are gf089 (early Eta) and gf085 (Meso/off-time Eta). Output grids are 211/Q (80 km grid), 212/R (40 km grid), and 215/U (20 km grid).

websites

weather.cod.edu/forecast
www.aos.wisc.edu/weatherdata/eta/
www.emc.ncep.noaa.gov/mmb/SREF/FCST/
    COM_US/web_js/html/eta12km_t2m.html
weather.unisys.com/eta

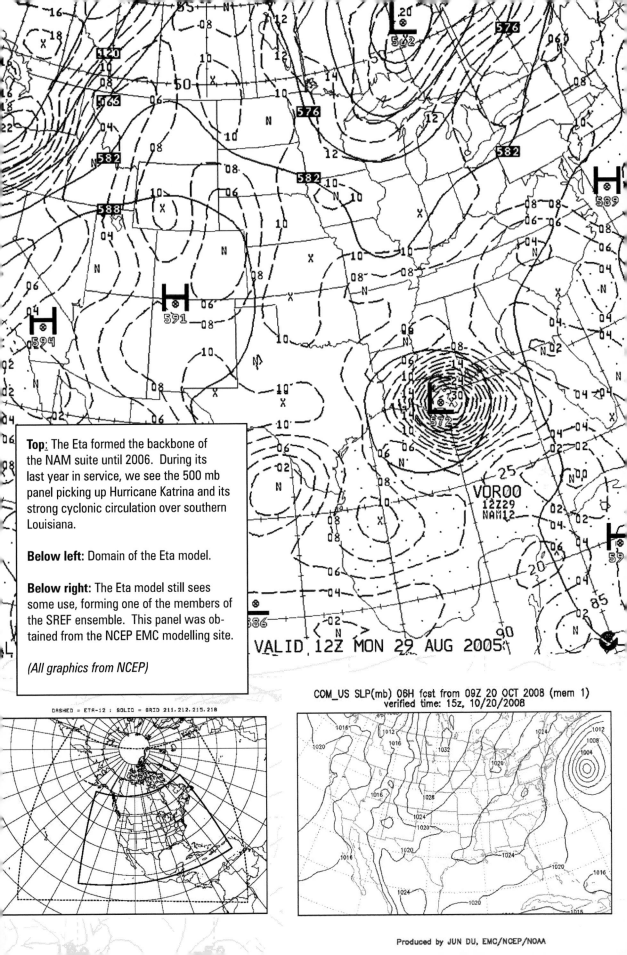

Top: The Eta formed the backbone of the NAM suite until 2006. During its last year in service, we see the 500 mb panel picking up Hurricane Katrina and its strong cyclonic circulation over southern Louisiana.

Below left: Domain of the Eta model.

Below right: The Eta model still sees some use, forming one of the members of the SREF ensemble. This panel was obtained from the NCEP EMC modelling site.

(All graphics from NCEP)

VALID 12Z MON 29 AUG 2005

DASHED = ETA-12 : SOLID = GRID 211, 212, 215, 218

COM_US SLP(mb) 06H fcst from 09Z 20 OCT 2008 (mem 1)
verified time: 15z, 10/20/2008

Produced by JUN DU, EMC/NCEP/NOAA

RUC model

During the late 1980s, ACARS aircraft observations, wind profiler data, and an increasingly dense surface network were all rapidly coming online. It became quite obvious that a new model was needed to synthesize all of this mesoscale data at a rapid rate and provide for frequent updates. There was also a need for a model that would end the long waits for major model runs to finish while weather changed by the hour. Thus, the RUC (Rapid Update Cycle) model was born.

The model was initially developed by the NOAA Forecast Systems Laboratory, where it was known as MAPS (Mesoscale Analysis and Prediction System). It underwent about three years of testing, and finally the completed model was implemented at the National Meteorological Center (NMC, now NCEP) in September 1994.

In April 1998 the RUC-2 model was introduced, replacing the original RUC. It added considerable improvements to the physics, parameterizations, and model resolution. GOES precipitable water data was introduced with this upgrade.

In April 2002, the RUC20 model was introduced, replacing the RUC2. It doubled the resolution and added 10 new vertical layers. There were also vast improvements to integration of surface data, improved diagnostic routines, better cloud parameterization, and inputs every 6 hours from the Eta run. Analysis was done through old-fashioned optimum interpolation (OI) up until sometime in 2003 when the 3DVAR method was integrated.

In June 2005 the resolution was increased to 13 km operationally. With this change it has begun using more input data, such as METAR cloud/visibility data and GOES cloud top data. In November 2008 another upgrade is expected which includes physics upgrades and the use of radar, mesonet winds, and aircraft ACARS data.

Model biases

The sigma-theta coordinate system delivers several advantages for the RUC model. First and foremost, it is theoretically much better at forecasting the onset of isentropic upglide compared to eta and pure sigma coordinate systems, however the RUC is theoretically worse than the older Eta model at handling mountain interactions such as cold air damming and lee cyclogenesis. The coordinate system also allows for better modelling of surface heating and dynamical mixing, as well as excellent physical representation of processes such as snow cover and evaporation.

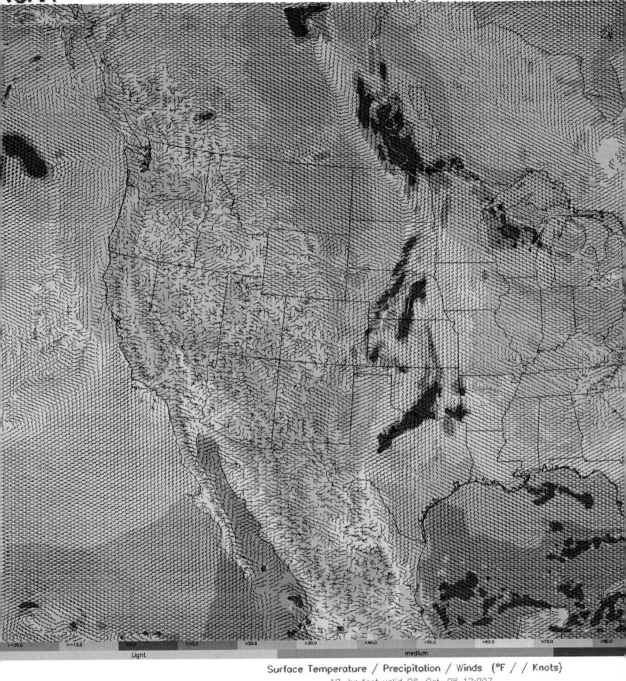

Surface Temperature / Precipitation / Winds (°F / / Knots)
12-hr fcst valid 06-Oct-08 12:00Z

Top: Forecast Systems Laboratory graphic of RUC
forecast output. It shows considerable small-scale
detail, particularly in the precipitation fields. *(FSL)*

Right: The detailed terrain field present in the RUC
model. The chart coverage corresponds to the actual
geographic coverage of the model. The RUC's sigma
surfaces "bend" over these terrain features. *(NCEP)*

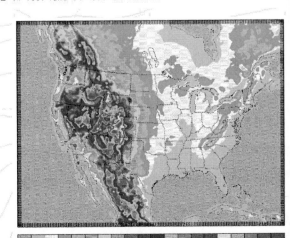

WRF

The Weather Research and Forecasting (WRF) model was completed under a partnership of U.S. government and academic users who aspired to emulate Europe's development of a robust "community" numerical model. The new model had to be flexible, portable, and scalable. A prototype was fielded in 2000. The model was implemented by National Centers for Environmental Prediction (NCEP) in June 2006, replacing the Eta in its NAM (North American Mesoscale) suite. It is expected to completely replace most of the NCEP legacy models.

The WRF is public domain, and binary and source code is freely available for download. It offers users a choice of algorithms from different contributors. The source code uses a layered software architecture that encapsulates low-level processing within the model to make custom routines flexible, independent, and efficient. The model runs grids smaller than 10 km, and is capable of these tight resolutions since it is a non-hydrostatic model. The WRF is designed for portability, with the ability to run under a number of architectures.

There are two primary versions of the WRF: the **ARW** (advanced research WRF; formerly called "EM" for Eulerian mass) maintained by the National Center for Atmospheric Research, and the **NMM** (non-hydrostatic mesoscale model) maintained by the National Centers for Environmental Prediction (NCEP). Each has a different set of equations at its core. Preliminary studies have found little skill difference between the ARW and NMM.

Starting in 2006, NCEP began using both the inhouse WRF/NMM and the NCEP WRF/ARW model with an ultimate goal of using it to replace several legacy models, including the Eta, RUC, and GFDL hurricane model. NCEP runs this model under a product suite known as NAM (North American Mesoscale). It can be thought of as an upgrade of the old NGM and Eta model runs.

Biases

Early on the NAM WRF was recognized for being aggressive with moisture, the opposite of the problem that used to plague the Eta model. The effect was particularly evident in broad forested regions such as the eastern United States. The parameterizations used for evapotranspiration were adjusted in 2007, alleviating this problem.

A 2008 study of turbulence found that the NAM tends to overforecast wind speeds in the upper troposphere, especially when associated with a subtropical jet with anticyclonic curvature.

a closer look

The WRF is a finite difference (gridpoint) model. It may use either eta or zeta vertical coordinate systems. The horizontal and vertical resolution is completely dependent on user specifications, and is often less than 10 km.

The model allows unlimited options for dynamics and physics packages, and can be tailored to a substantial degree. This allows the end user to select the ideal configuration for a given purpose.

The WRF, taking on dual roles as a mesoscale and regional model, is expected to replace all NCEP model runs.

The WRF will support nested grids, which was the concept of the NGM model in the 1980s.

The first release of the WRF was on November 30, 2000. Real-time forecast grids began trickling out from NCAR in March 2001 and from NCEP in 2006.

Future upgrades are expected to include input of actual observed data from the MM5 model, output in other formats besides netCDF, new microphysics processes, and more.

It is possible for anyone to run the WRF at home or work if they have a Linux computer and some Linux know-how. On a typical PC a state-size 10 km model can be run in about 1.5 hours.

Full information about the WRF model can be found at <www.wrf-model.org>

websites

www.nco.ncep.noaa.gov/pmb/nwprod/analysis
www.nssl.noaa.gov/wrf
www.ilmeteo.it/portale/mappe-meteo-wrf
box.mmm.ucar.edu/wrf/REAL_TIME

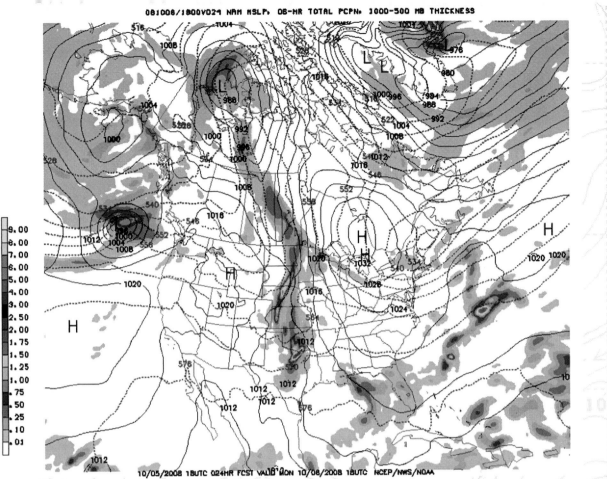

081006/1800V024 NAM MSLP, 06-HR TOTAL PCPN, 1000-500 MB THICKNESS

10/05/2008 18UTC 024HR FCST VALID MON 10/06/2008 18UTC NCEP/NWS/NOAA

Above: WRF/NMM forecast of sea level pressure, 1000-500 mb thickness, and precipitation as produced by the NCEP NAM run. The WRF replaced the Eta in 2006. *(NCEP)*

Right: WRF/ARW 8 km run performed by author at home on a Celeron 2.2 GHz Linux machine. Execution time for a 18-hour forecast was about 1 hour. This is a testament to the portability and flexibility of the WRF architecture.

Dataset: WRF RIP: sfc-wind
Fcst: 4.00 h
Horizontal wind speed
Horizontal wind vectors
Sea-level pressure

Init: 0600 UTC Thu 16 Oct 08
Valid: 1000 UTC Thu 16 Oct 08 (0500 CDT Thu 16 Oct 08)
   at height = 0.01 km
   at height = 0.01 km

sm= 2

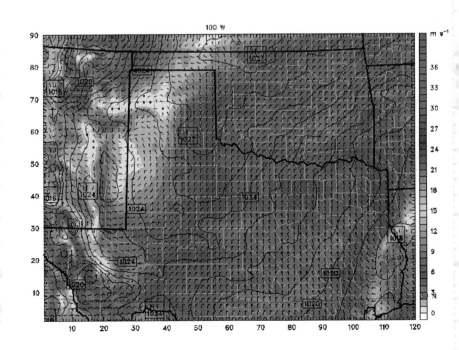

CONTOURS: UNITS=hPa LOW= 1017.0   HIGH= 1027.0   INTERVAL= 1.0000
BARB VECTORS: FULL BARB = 5 m s⁻¹
Model Info: V2.1.2   YSU PBL WSM 6class Noah LSM 10 km,   30 levels,   60 sec
LW: RRTM SW: Dudhia DIFF: simple KM: 2D Smagor

GFS model

The term GFS (Global Forecast System) is the United State's global forecasting model and the counterpart to the European ECMWF model. It encompasses a broad suite of products that includes the AVN (Aviation Model), the GDAS (Global Data Assimilation System), and the discontinued MRF (Medium Range Forecast) model.

The model was first introduced operationally as the GSM (Global Spectral Model) on 18 March 1981, replacing the PE (Primitive Equation) model that had been operational since 1966. It was the first operational model which visualized the atmosphere as mathematical waves rather than arrays of data at gridpoints. It started out as a 30-wave 12-level model, in other words, weather systems were defined as a series of 30 waves around the hemisphere, and 12 horizontal surfaces were used to represent the depth of the troposphere. The 30 waves gave it an effective resolution of 465 km.

In 1985, the GSM was split into two runs: the AVN (Aviation) for short-range forecasting and the MRF (Medium Range Forecast) model for medium-range forecasting. The models were identical, but the MRF was started several hours later than the AVN to make full use of all available data. On 23 April 2002, the MRF and AVN were consolidated back into one run, called the Global Forecast System (GFS), a suite designed to give the maximum view of global weather using all available data.

Model biases

A USGS user in a tip to NCEP suggested that the GFS was too ambitious with the strength and speed of systems crossing the Sierra Nevada range after 36 hours, and that the model terrain could be at fault.

The GFS has occasional problems with "precipitation bombs": a spurious convective complex with no apparent source. For more information on this effect see <*www.hpc.ncep.noaa.gov/qpfbombs*>.

In meridional flow the GFS can be too aggressive with amplifying the pattern and driving cold outbreaks southward, particularly past 72 hours. It is a good idea to consider other guidance such the ECMWF.

In split flow patterns, particularly during the cool season, the GFS has a tendency to phase the waves in each jet into a large long wave. This may overstrengthen surface systems on the High Plains of the United States. The UKMET model has been noted as a better choice in such situations.

a closer look

The GFS is a 382-wave spectral model with triangular truncation. This is a resolution of about 35 km. At the 180-hour point it becomes a 190-wave model. The GFS uses 64 pure sigma (terrain-following) levels. Time integration is leapfrog and semi-implicit.

Although the GFS is a spectral model, it offloads fields to a 768 x 384 point Gaussian grid for physics and nonlinear calculations.

It takes NCEP's IBM supercomputers about 12 minutes of computation time to complete the entire GFS run. However a considerable amount of time is spent preparing the data.

The GFS, being a sigma model, is prone to errors in the lee of mountain ranges. It has inherent problems with lee-side weather systems and cold-air damming. However it has excellent resolution near the ground and will accurately handle diurnal heating and low-level winds and moisture.

The GFS is in widespread use for world-wide forecasting, even in Europe and Asia, since it is one of the only full-scale planetary weather models whose output is disseminated in its entirety without restriction. Only limited portions of other global models are available for public consumption.

The GFS is run four times a day (00Z, 06Z, 12Z, and 18Z) out to 384 hours.

Though output is available through 16 days into the future, skill deteriorates significantly beyond 5 to 7 days and errors tend to increase exponentially.

websites

www.nco.ncep.noaa.gov/pmb/nwprod/analysis
www.emc.ncep.noaa.gov/modelinfo/
wxweb.meteostar.com

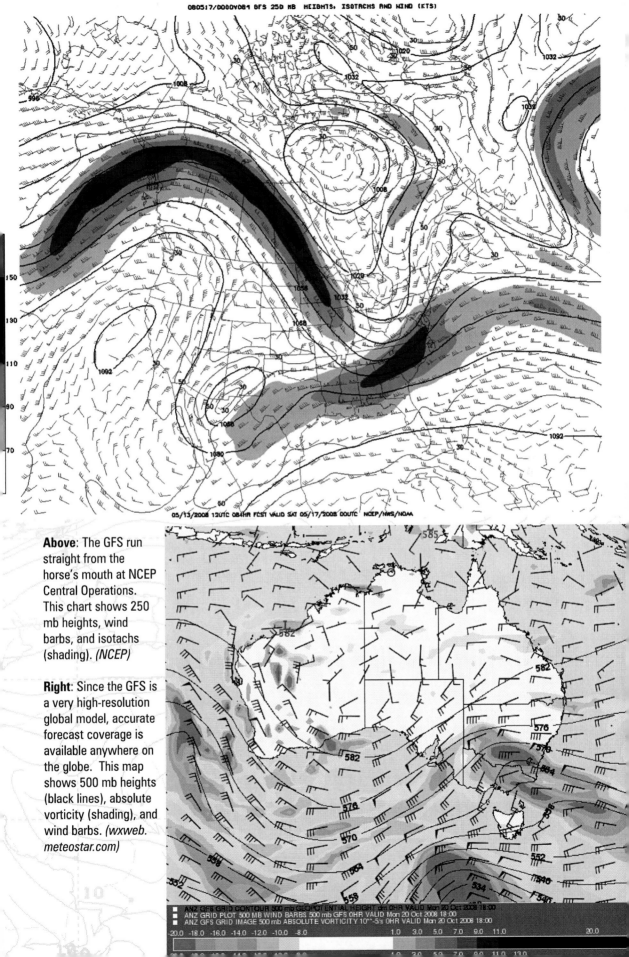

Above: The GFS run straight from the horse's mouth at NCEP Central Operations. This chart shows 250 mb heights, wind barbs, and isotachs (shading). *(NCEP)*

Right: Since the GFS is a very high-resolution global model, accurate forecast coverage is available anywhere on the globe. This map shows 500 mb heights (black lines), absolute vorticity (shading), and wind barbs. *(wxweb. meteostar.com)*

MM5 model

The MM5 is not the name for a suite of products but rather for a "brand" of model, the Mesoscale Model Fifth Generation developed by the University Corporation for Atmospheric Research (UCAR) in collaboration with Pennsylvania State University (PSU). It was designed to be the ultimate solution to a pressing requirement in the research community: the need for a small-scale model that was not locked onto a large, regional domain.

The model was first developed as early as 1972, and as it came to the forefront of mesoscale meteorology with the arrival of powerful computing technology, the MM5 was released to the weather community in early 1994. The first generation of the model was optimized for Cray systems. The second version, released the same year, supported more operating systems and included improved physics packages.

The days of the MM5, however, are numbered. The WRF (q.v.) is expected to quickly replace it as the "community" model of choice due to its newer, more robust design and flexible architecture. The U.S. Air Force, which was one of the biggest users of the MM5 during the late 1990s and early 2000s, replaced all of its MM5 windows with WRF/ARW models in late 2007.

Since it is adapted for high-resolution terrain and uses non-hydrostatic calculations, the MM5 is well suited for modelling extremely small-scale processes. It can be used to examine mesoscale convective systems, fronts, tertiary circulations, and urban meteorology effects. Furthermore its design for portability allows it to be run on a broad range of Unix systems, with vast flexibility on the resolution, the domain size, and other configuration options. The source code is in FORTRAN and can be easily modified to customize the model.

The AFWA configuration of the MM5 was noted in 2000 by NCEP to have problems with overforecast rain totals in the southeast and eastern U.S., particularly in regions of synoptic-scale forcing. The initial analysis scheme was suspected to be partly to blame.

Unfortunately it is impossible to give a complete description of MM5 model biases since it is run by different users in a variety of configurations and with varying data sources. Very little in the way of quantitative studies of model performance have been done. However the consensus is that the MM5 does an excellent job with mesoscale weather systems, true to its design.

a closer look

The MM5 model is a finite difference (gridpoint) model with pure sigma (terrain-following) surfaces. The horizontal and vertical resolution is set by the user.

The model is non-hydrostatic, which means that it can be run at unusually small scales (with resolutions less than 10 km).

The MM5, being a sigma model, is prone to errors in the lee of mountain ranges. It will have problems with lee-side weather systems and cold-air damming. However it has excellent resolution near the ground and will accurately handle diurnal heating and low-level winds and moisture.

In 2003, the MM5 was in operational forecasting use in exotic locations like Kenya, Mozambique, China, Peru, and Colombia.

Is there an MM4? Yes, there is. Papers were published on this preliminary version in 1987 by Hsie and Anthes.

Model milestones:
Initial release: February 1994
Version 2 (V2): July 1994
Version 3 (V3): June 1999

websites

www.mmm.ucar.edu/prod/rt/pages/simple.html
https://afweather.afwa.af.mil/weather/met/
    met_home.html
www.atmos.washington.edu/mm5rt
aurora.aos.wisc.edu/current.shtml
met.nps.edu/~hale/MM5
cheget.msrc.sunysb.edu/html/alt_mm5.cgi
galileo.imta.mx/simumm5.php
www.mmm.ucar.edu/mm5

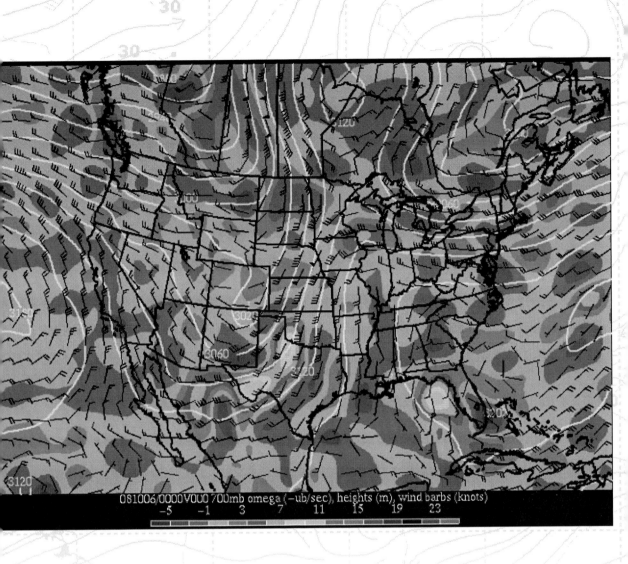

081006/0000V000 700mb omega (−ub/sec), heights (m), wind barbs (knots)
−5   −1   3   7   11   15   19   23

Above: MM5 prediction for the United States generated at the University of Wisconsin.

Right: MM5 output for the Los Angeles area, coastal range, and southern San Joaquin Valley. *(Navy/NPS)*

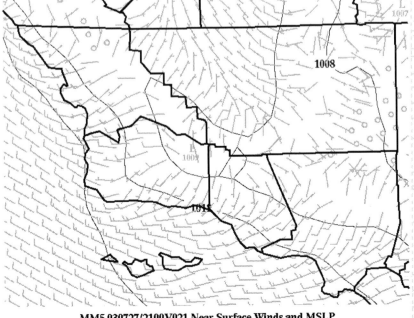

MM5 030727/2100V021 **Near Surface Winds and MSLP**

NOGAPS

The NOGAPS (Navy Operational Global Atmospheric Prediction System) is a global spectral model that runs twice a day out to 144 hours. It is managed by the U.S. Navy's Fleet Numerical Meteorology Center (FNMOC). The NOGAPS model uses the multivariate version of the OI (optimal interpolation) analysis scheme. It executes on SGI Origin 3000 supercomputers at FNMOC in Monterey, California.

The Navy had been experimenting with numerical models as early as 1959. In the mid-1970s there was a push for a global spectral model to be developed, and the Naval Research Laboratory produced the first version of NOGAPS in 1982. It was a finite-difference (gridpoint) model. Unfortunately during long-term testing it was found that NOGAPS was a poor contender to the AVN/MRF and ECMWF models. NOGAPS was scrapped in 1987 and rebuilt as a spectral model, yielding a much more robust system.

The model was good enough to become operational on January 1988 and has gone through a series of upgrades. The current NOGAPS release, V4.0, was fielded on 18 September 2002, and is run on SGI Origin 3000 supercomputers.

Model biases

Mature cyclones over land tend to be too deep during later time frames, generally past 2-3 days, while those over the ocean tend to be too weak. The best skill is with deepening lows over North America, but filling is too slow.

Deepening cyclones are moved too slowly over the ocean during meridional flow, and filling ones are moved too far poleward.

Deepening cyclones are moved too fast over the ocean during zonal flow, and filling ones are moved too far poleward.

Surface and upper-level cyclones are deepened too slowly equatorward of the polar front jet, while those poleward are deepened too rapidly.

Oceanic anticyclones are slightly too intense.

The model does a good job handling transitions from digging troughs into cutoff lows in the spring and fall.

Upper-level troughs are dug too aggressively during the cool season, on the U.S. West Coast beyond 84 hours.

The model is slightly overprogressive with upper troughs in zonal flow.

Tropical cyclones tend to be moved too slowly.

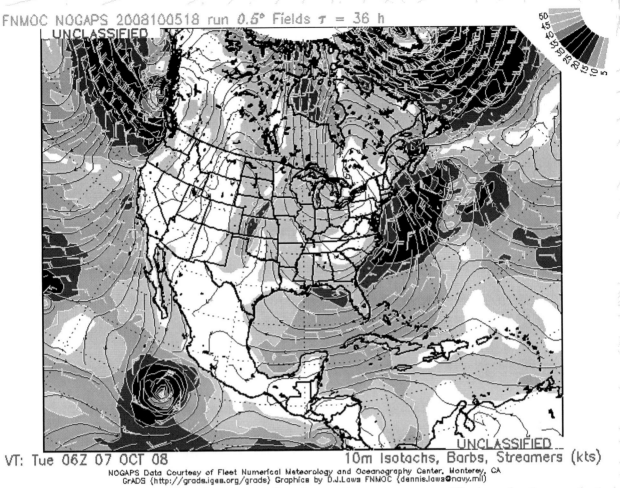

FNMOC NOGAPS 2008100518 run 0.5° Fields τ = 36 h
UNCLASSIFIED

UNCLASSIFIED

VT: Tue 06Z 07 OCT 08   10m Isotachs, Barbs, Streamers (kts)

NOGAPS Data Courtesy of Fleet Numerical Meteorology and Oceanography Center, Monterey, CA
GrADS (http://grads.iges.org/grads) Graphics by D.J.Laws FNMOC (dennis.laws@navy.mil)

Above: NOGAPS forecast as produced by the U.S. Navy FNMOC web site. *(FNMOC)*

Right: Due to the limited availability of ECMWF fields, the NOGAPS model has proven to be a valuable forecasting tool in Europe. This depiction is available at <www.wetterzentrale.de>. 500 mb contours are drawn as shading. A frontal system is moving through the North Sea, France, and the western Mediterranean Sea, with deep upper troughing over the Bay of Biscay. *(Wetterzentrale)*

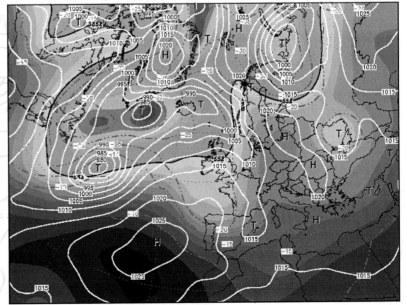

Init : Mon,06OCT2008 18Z   Valid: Wed,08OCT2008 06Z
500 hPa Geopot.(gpdm), T (C) und Bodendr. (hPa)

Daten: NOGAPS-Modell der US-Navy
(C) Wetterzentrale
www.wetterzentrale.de

ECMWF

The ECMWF is the name for a meteorological center as well as a model: The European Centre for Medium-Range Weather Forecasts. The center was envisioned in 1967 as part of a resolution to develop a multinational weather center, and opened its doors in 1973.

The first operational ECMWF model forecast was produced on 1 August 1979 using a Cray 1A supercomputer, about as powerful as today's home computers. It was a 15-level gridpoint model with a resolution of 200 km. In April 1983, ECMWF adopted a spectral model with 63 waves and 16 layers. In 1992 an ensemble forecasting system was added. In April 1995 parameterization of clouds was added. In 1996 the Optimum Interpolation analysis scheme was replaced by the 3DVAR method. The WAM ocean wave model was introduced in 1992, followed by integration with the atmospheric model in 1998.

As of 2008, the ECMWF model is run on an IBM Cluster 1600 system, featuring a scalar rather than vector architecture. It consists of 155 p5 575 servers, similar to the configuration at the United States NCEP. The IBM system produced its first ECMWF forecast in March 2003. It has a combined RAM of 2.25 TB and about 50 TB of hard drive storage.

Model biases

The ECMWF is considered to be superior at forecasting upper-level heights during the cold season, particularly with respect to wave number transitions and the onset of +PNA circulation episodes (west Canada ridge with polar air affecting the central and eastern U.S.).

The model is excellent at handling timing of shallow cold air outbreaks, particularly in the Great Plains.

The ECMWF is notorious for overdeveloping or overpopulating cutoff lows, particularly in the southwestern U.S. However this yields somewhat better skill than other models in spring when cutoff lows are most common. The model is also too slow or even retrogressive with cutoff lows, which are sometimes even erroneously shunted westward underneath the Pacific subtropical high.

High height bias in the upper troposphere is a well-documented problem, producing a high thickness bias.

The model has a meridional bias, making upper air patterns unusually amplified and surface systems more intense and more slow. Therefore a progressive pattern forecast by the ECMWF is significant and probably meaningful.

a closer look

The ECMWF model is a 799-wave spectral model (about 25 km resolution) with triangular truncation. It uses 91 hybrid sigma-pressure levels, consisting of sigma (terrain-following) surfaces near the ground and pressure surfaces aloft, and moves forward in 12-minute time steps.

The ECMWF model must make 1.63 quadrillion calculations in order to make one complete forecast.

A considerable amount of data is restricted from the public, in particular model fields before the 72 hour point. This has been a longtime issue of debate between the European private-commercial sectors and the academic-government sectors. Even European hobbyists must turn to the NOGAPS and GFS to examine detailed products.

In 1997 the ECMWF was the first model to use the 4DVAR scheme for data input, which takes into account temporal changes in the data.

Like most global models, the ECMWF ingests surface observations, radiosonde, aircraft ACARS, and geostationary satellite wind data.

This model is supported by 28 European member states, and is generally considered to be the most sophisticated and computationally expensive numerical prediction model used for global forecasting. The operations budget of the ECMWF data center is about 40 million Euros.

websites

weather.cod.edu/forecast
www.wetterzentrale.de/topkarten
www.wxcaster.com/
   conus_0012_foreign_models.htm
http://www.ecmwf.int/products/forecasts/guide/
weather.unisys.com/ecmwf

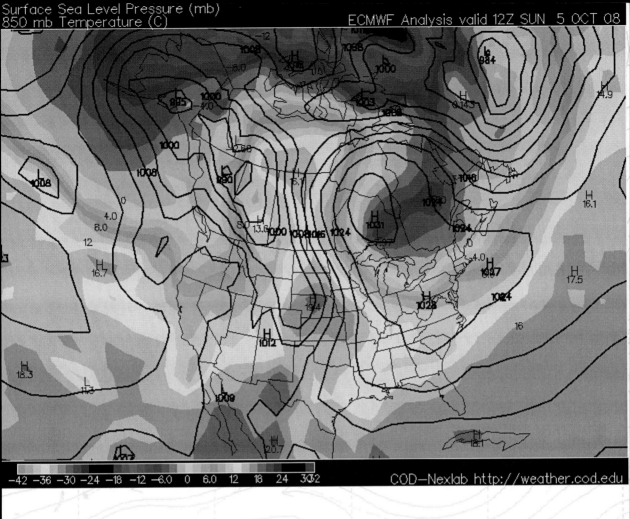

-42 -36 -30 -24 -18 -12 -6.0 0 6.0 12 18 24 30 32

COD—Nexlab http://weather.cod.edu

Above: ECMWF output as provided on the Unisys site <weather.unisys.com>. *(Unisys)*

Right: The ECMWF supercomputer facility near Reading, England consists of IBM p5 575 supercomputers driven by the AIX operating system. Each cluster contains 2.25 terabytes of memory and 100 terabytes of disk storage. In a single second it can make calculations that would take 17 million years for a person to do by hand. Each of the 155 servers can draw as much as 3.4 kW, which is equivalent to the power consumption of a small household central air conditioning unit. *(ECMWF)*

UKMET

The term UKMET is a shorthand phrase describing the Unified Model (UM) system, a sophisticated set of numerical forecast models operated by the UK Met Office. It consists of three sets of model runs: the Global, the NAE, and the UK mesoscale window. At press time, the latest incarnation of the UM was 6.3.

Great Britain's expertise with numerical models goes back to 1959, when the University of Manchester successfully ran a numerical prediction on a Ferranti Mark 1 computer. Almost immediately the British Met Office put a Ferranti Mercury computer to work, producing experimental 36-hour forecasts for the eastern Atlantic and western Europe. It took about 6 hours to produce each complete forecast.

With a new computer purchase in 1965, the Met Office introduced its first operational model that covered 30 hours and up to 72 hours experimentally. This was improved further with an IBM 360 acquisition in 1972 and introduction of a nested grid scheme. The model has undergone upgrades since then, the latest being the addition of a new physics package nicknamed "New Dynamics" on 7 August 2002. The model was moved to new NEC SX-6 systems in March 2004.

The UM system has been run for decades at the Bracknell facility west of London, but during the summer of 2003 the facility was moved to the Met Office's new headquarters at Exeter in Devon.

Due to the rapid upgrades of the UM during the 2000s, very little applicable information has been published on qualitative aspects of the model's performance. Biases historically associated with the UKMET, but which may not be representative anymore, include:
- A persistent warm bias in the middle troposphere
- Problems with shallow cold air outbreaks
- A zonal bias; i.e. a refusal to amplify long wave troughs
- An equatorward bias for surface and upper-air systems.
- An equatorward bias for the polar jet and prevailing westerlies
- Short waves which move too fast.

a closer look

The Met Office Unified Model System ("UKMET") is a finite difference (gridpoint) model which uses nested grids and spherical coordinates.

The Global Forecast model features 40 km resolution with 70 hybrid levels.

The NAE (North Atlantic and European) model features 12 km resolution with 70 hybrid levels.

The UK Mesoscale model features 4 km resolution with 70 hybrid levels.

The hybrid levels are sigma-pressure levels, with sigma (terrain-following) coordinates close to the ground and pressure surfaces aloft.

The UKMET hybrid vertical coordinate scheme is also used by the ECMWF, NOGAPS, and JMA models.

An NCEP bias report suggests that the UKMET performs better than the GFS with phasing of weather systems in the northern and southern branch of the jet, particularly in North America. It suggests when this configuration exists beyond 84 hours that the UKMET output should be checked.

The UKMET model consists of 900,000 lines of portable Fortran 77/90 and C source code.

websites

cumulus.geol.iastate.edu/ukmet.html
twister.sbs.ohio-state.edu/models/ukmet
meteocentre.com/models/modelsukmet_e.html
www.wetterzentrale.de/topkarten/fsukmeur.html
www.wxcaster.com/
  conus_0012_foreign_models.htm
www.meto.govt.uk/research/nwp/numerical/
operational

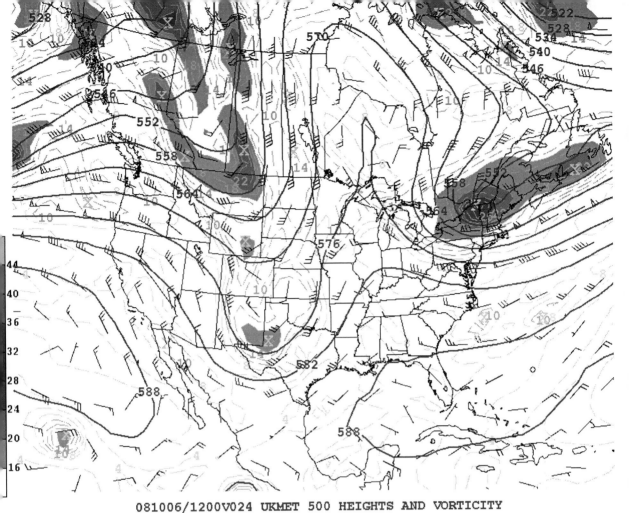

081006/1200V024 UKMET 500 HEIGHTS AND VORTICITY

PN 0 0 Pressure hPa
D5 0 0 1000-500 hPa Thickness dam
GZ 500 0 Geopotential Height dam

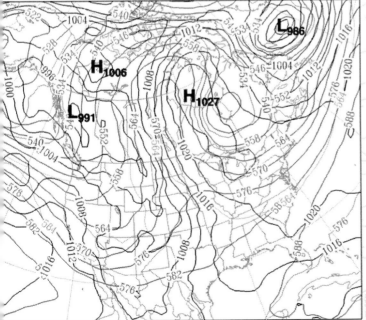

Valid at Sun Oct 5 00:00:00 2008
UKMET 00Z MWC http://meteocentre.com/

Above: The Iowa State weather server provides excellent UKMET maps of North America.

Left: ECMWF from Meteocentre.com.

Below: The nested mesoscale subset of the Unified Model System covers Great Britain at 17 km resolution. The remainder of the globe is covered at 62 km resolution.

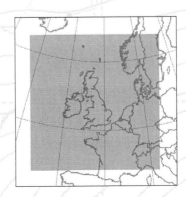

GEM

The term GEM (Global Environmental Multiscale) actually refers to two main types of models: the Global (formerly known as the SEF) and Regional (formerly known as the regional fine-elements model: RFE, or EFR). The GEM is a global spectral system run at the CMC computing facility in Dorval, Quebec.

The GEM Global model forecasts out to 72 hours (12Z base time), 240 hours (00Z base time), and 360 hours (Saturdays). It uses the 3DVAR analysis scheme.

The GEM Regional model is a variable-gridpoint model that produces output from a base time of 00Z and 12Z, forecasting out to 48 hours. Operating with 15 km resolution and 58 hybrid sigma-Z levels, it uses the Kain-Fritsch deep convection scheme and the Kuo transient scheme for shallow convection.

Canada's technological developments began in 1950 with the introduction of an experimental barotropic model running on an IBM 650 at McGill University. Operational charts did not begin until July 1963 when a barotropic model began producing output for 500 mb.

On 18 February 1976 the first global spectral model was implemented. It featured 20 hemispheric waves with rhomboidal truncation and five vertical levels. In 1987 CMC acquired a Cray X/MP 28 for its model suite.

On 22 April 1986, the first regional model, analogous to the NGM and Eta, was introduced. It was called the RFE (Regional Finite Element), also known as EFR (Éléments Finis Régionaux) and featured 15 layers with 190 km resolution, and was placed on the Cray X/MP in 1987 with better resolution.

On 24 February 1997 the RFE became the GEM Regional model, and on 14 October 1998 the SEF was decommissioned. It was replaced by the GEM Global model, a 0.9-deg global uniform grid model with 28 eta levels.

The GEM has proven to be an excellent, accurate model. Since it sigma coordinates, it does have difficulty handling lee-mountain effects such as lee cyclogenesis and cold air damming. The GEM model has also been faulted in rejecting too much upper air data which appear to be valid but don't fit the first-guess fields.

The regional model is considered to be a very accurate forecast system. It is known to have a cold bias at the surface during wintertime nocturnal cooling episodes owing to the model's deficit in downward long-wave radiation. The model also tends to overforecast precipitation on the windward side of mountains, especially during the spinup cycle (roughly the first 6 hours).

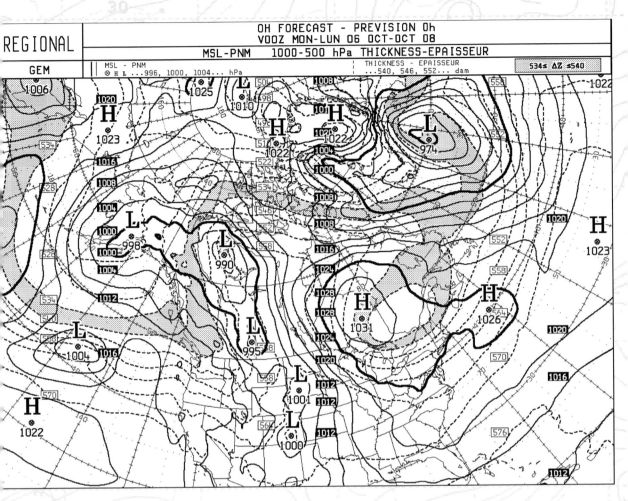

REGIONAL

GEM

OH FORECAST - PREVISION 0h
V00Z MON-LUN 06 OCT-OCT 08
MSL-PNM   1000-500 hPa THICKNESS-EPAISSEUR

MSL - PNM
⊗ H L ...996, 1000, 1004... hPa

THICKNESS - EPAISSEUR
...540, 546, 552... dam

534≤ ΔZ ≤540

Above: Environment Canada provides an excellent counterpart to the legacy NCEP DIFAX style at <weatheroffice.ec.gc.ca>. *(CMC)*

Right: The global grid of the GEM Global Spectral Model, with the nested domain that makes up the Regional Spectral Model. This smaller grid covers all of the United States except south Texas and Hawaii. *(CMC)*

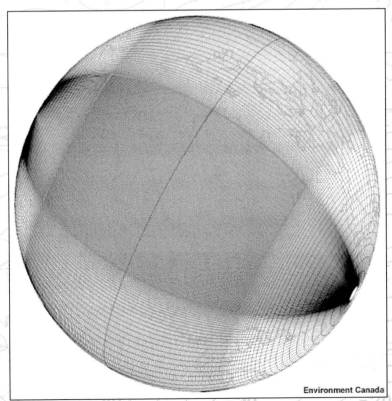

Environment Canada

JMA

The term JMA in American forecasting use usually refers to the global spectral model (JMA/GSM), which yields worldwide coverage. Japan also operates five other models including a regional spectral model (JMA/RSM), a mesoscale model (JMA/MSM), and a typhoon model (JMA/TYM).

Japan began its numerical weather prediction efforts in 1959. The first global-scale spectral model came online in March 1988 using sigma coordinates, but this was rapidly changed to hybrid coordinates in November 1989. Hybrid coordinates follow terrain near the surface, gradually following pressure surfaces in the upper troposphere. Since then, the GSM has undergone periodic upgrades in its physics package and resolution.

In November 2007, the RSM and TYM were retired as their capabilities had been taken over by the MSM mesoscale model, which had undergone significant improvements that year.

As of 2008, the JMA numerical prediction suite consists of a GSM (global), MSM (mesoscale), and an ensemble prediction system.

The GSM model is a spectral global model running at a truncated 959 waves with 60 sigma-p hybrid pressure surfaces. It uses the Arawaka-Schubert scheme for cumulus parameterization, with the Mellor-Yamada Level 2 scheme to model the boundary layer.

The MSM model is a 721 x 577 gridpoint model centered on eastern Asia, with a resolution of 5 km and 50 hybrid terrain-following surfaces. It uses the modified Fritsch-Kain scheme for cumulus parameterization and, like the GSM, the Mellor-Yamada Level 2 scheme for boundary layer processes.

The model is run on Japanese supercomputers, notably the NEC during the early 1990s and Hitachi systems since 1993.

a closer look

The JMA global spectral model became a 959-wave spectral model with triangular truncation, effective with the November 2007 update, yielding a resolution of 20 km. It uses 60 hybrid sigma-pressure levels, with sigma (terrain-following) surfaces near the ground changing to pressure surfaces above the middle troposphere.

The JMA mesoscale spectral model (MSM) features a resolution of 5 km and uses 50 hybrid layers. The model calculates out as much as 33 hours in the future.

The model uses a hybrid vertical coordinate scheme, which is also used by the ECMWF, NOGAPS, and UKMET models.

websites

gpvjma.ccs.hpcc.jp/~gpvjma/weathermap.html
www.wxcaster.com/
  conus_0012_foreign_models.htm
www.wright-weather.com (Pay)
www.wetterzentrale.de/topkarten/fsjmaeur.html

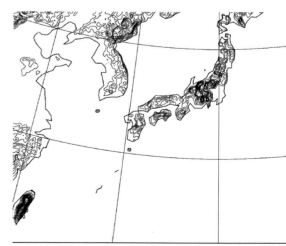

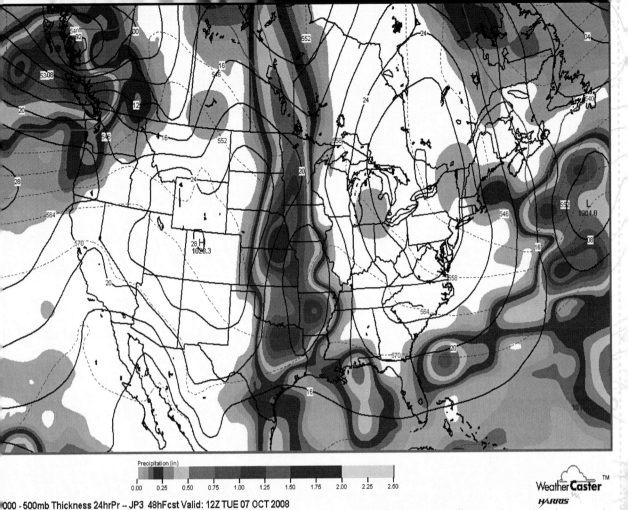

Precipitation (in)

0.00 0.25 0.50 0.75 1.00 1.25 1.50 1.75 2.00 2.25 2.50

000 - 500mb Thickness 24hrPr -- JP3 48hFcst Valid: 12Z TUE 07 OCT 2008

Weather**Caster** ™

HARRIS

Above: Output for the United States from the Japanese JMA Global Spectral Model (JMA/GSM) as obtained at Earl Barker's model site. Output is generally available from 00 to 168 hours for North America. Sources for other regions are not known. *(wxcaster.com)*

Right: Many of the real-time output fields are available on the GPV/JMA Archive site. The resolution is sorely lacking, however it can provide a quick overview of medium-range patterns. *(University of Tsukuba/CCS)*

Opposite: The domain and topography of the JMA Mesoscale Model (JMA/MSM). *(JMA)*

500 hPa Height

2008 1020 12Z H+000 JMA/GPV/GSM

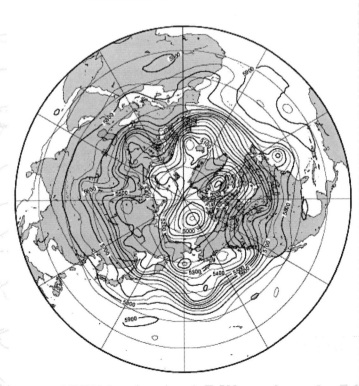

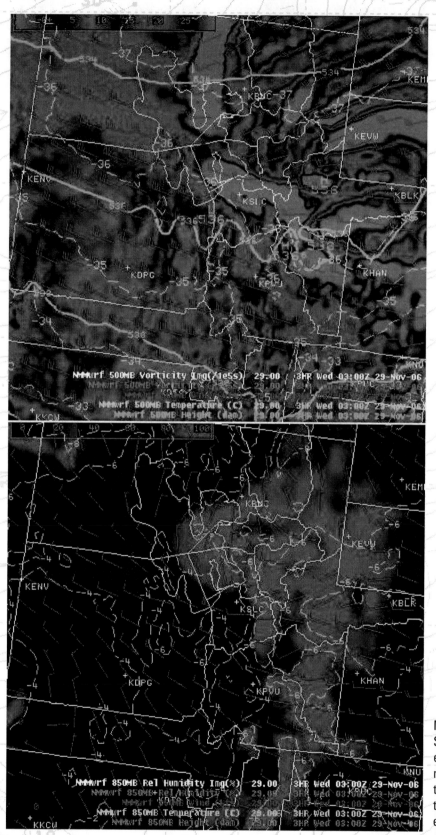

National Weather Service forecasters view numerical model output using the AWIPS workstation, as shown by the examples here.

CHAPTER 6
RAW DATA

SYNOP surface data

The SYNOP format is given in WMO Pub. 306, "Manual on Codes" in section FM-12. It specifies the format that must be used worldwide for all non-aviation weather observations (aviation reports are coded under METAR format).

The format is as follows:

AAXX hhmmw

CCCCC rxhvv NDDFF 1sttt 2SDDD 3pppp 4PPPP 5appp 6RRRT

7wwpp 8NLMH 9hhmm

If any particular group is missing, it is assumed the category is not applicable. For example, if the sky is clear, the 8NLMH cloud code group will be omitted. Solidii ("/") usually indicate missing data.

Header (AAXX hhmmw). The header usually appears at the top of a group of reports collected by a single office. It always starts with "AAXX" ("BBXX" indicates a ship report with more elaborate coding standards that are beyond the scope of this book). The time follows in hours (hh) and minutes (mm). If the wind indicator (w) is 0 or 1, winds are in m/s; if 2 or 3 winds are in knots; an even value is estimated and odd is measured.

Station location (CCCCC). This represents the five-digit WMO identifier where the weather was observed.

Miscellaneous and visibility group (rxhvv). The precipitation indicator (r) tells whether a supplementary block is used. Station type (x) is 1-3 if manned, 4-7 if automated. Lowest cloud height (h) is a coded value. Visibility (vv) is also a coded value; when below 50 it generally indicates the visibility in tens of kilometers.

Wind group (NDDFF). This group begins with the total cloud cover (N) in eighths; if it is 9 the sky is obscured and if a solidus it is not known. The wind direction in tens of degrees relative to true north (dd) and speed (ff) are given. The units are given in the header (see above).

Temperature group (1sttt). Exact temperature (ttt) in tens of degrees Celsius. If the sign value (s) is "1", the temperature value is negative.

Dewpoint group (2SDDD). Exact dewpoint temperature (DDD) in tens of degrees Celsius.

If the sign value (S) is "1", the dewpoint value is negative.

Station pressure (3pppp). The station pressure (pppp) is in tens of millibars.

Sea-level pressure (4PPPP). The sea-level pressure (pppp) is in tens of millibars. If the first digit is 1, 2, 5, 7, or 8 this indicates that this is not a sea-level pressure and is instead a geopotential height value.

Sea-level pressure (5appp). Pressure tendency (a) with 0-3 risen, 4 steady, and 5-8 fallen, and change (ppp) in tens of millibars.

Precipitation (6RRRT). Liquid precipitation amount (RRR) in whole mm. "990" is a trace, and with higher values the last digit is the value in tenths of a mm. The duration of the period (T) is "4" if 24 hours, "2" if 12 hours, and "1" if 6 hours.

Weather (7wwpp). The current weather (ww) and past weather (pp) is expressed as a two-digit code (see code table). In general the higher the number the more significant the phenomenon.

Cloud group (8NLMH). The amount of low or middle clouds (N) is in eighths. The code for any low (L), middle (M), or high (H) clouds is given.

Time group (9hhmm). The observation time is given in hours (hh) and minutes (mm) UTC.

409
SMRS15 RUMS 241800
AAXX 24181
22282 32598 71803 10134 20113 40097 56002 85214   333 10163=
22292 32473 61303 10196 20166 40081 56001 85930   333 10241=
22365 NIL=
22438 32571 71802 10182 20165 40105 53003 82274   333 10226=
22446 NIL=
22563 32882 72102 10220 20173 40113 53008 86083   333 10278=
22641 11684 32101 10220 20175 40121 53002 69900 70281 81242   333 10265=
22695 NIL=
22778 11472 79901 10193 20182 40152 52007 69940 70298 82461
333 10274=
22798 12984 71802 10208 20174 40158 54000 69900 85051   333 10275=
22854 11884 31702 10216 20156 40148 57002 60010 70198 81031   333
10236=
22867 11971 21801 10216 20187 40160 57001 69900 70281 80001   333
10242=
22939 32697 21503 10212 20148 40158 53001 81501 333 10272=
22996 12598 60201 10210 20159 40176 53004 69900 83905 333 10280=
27008 32582 72302 10235 20161 40156 57001 87900 333 10284
85920=
27051 32960 82301 10212 20126 40175 53001 80001 333 10268=
27225 33568 22003 10208 20133 40186 53001 82101   333 10262=
27252 NIL=
27369 32581 72702 10222 20158 40191 52004 87300   333 10274 87925=
27393 32598 20000 10224 20144 40189 53001 81401   333 10279=
27479 32599 32401 10225 20153 40193 53006 83200   333 10276=
27532 32997 30000 10198 20149 40193 56001 82031   333 10252=
27648 32997 20000 10192 20155 40203 53002 82030   333 10268=
27679 32980 22302 10232 20100 40198 57006 82040   333 10270=
28214 32996 20000 10203 20141 40192 53003 80008   333 10277=

SMCI07 BABJ 241800 RRA
AAXX 24181
51716 31958 23605 10244 20105 38809 49994 52028 70600 80002
333 00556 10344=
51765 32980 70000 10224 20123 39071 49987 52008 82032 333
00253 10342=
51777 32968 00703 10276 20056 39027 49971 52011 333 00300
10358=
53149 32680 12702 10175 20166 38675 49986 54000 81500 333
00055 10279=
53231 32980 02204 10177 20092 38402 52002 333 00357 10257=
53336 32980 00000 10206 20105 38623 49992 52010 333 00000
10276=

METAR surface data

METAR stands for Meteorological Airport Report, and is the worldwide standard for transmitting weather reports from airfields. It is the backbone of weather reports in the United States, Europe, and the Pacific Rim.

Familiarity with METAR format is important for a forecaster to be able to pick up on minor trends that might occur at a weather station. The format is:

CCCC ddhhmmZ (AUTO) dddff VV ww CCCHHH tt/dd P (RMK)

Station location (CCCC). This represents the four-letter ICAO identifier where the weather was observed.

Observation time (ddhhmmZ). The UTC time the observation was taken: calendar day (dd); hour (hh); and minute (mm). The "Z" ending is a reminder that the time zone is Zulu (UTC) time.

Auto flag (AUTO). If this flag is present, it indicates the observation was taken by a machine.

Wind (dddff). The wind direction in degrees relative to true north (ddd) and speed (ff). If winds are gusting, the group takes the form dddffGgg, where gg is the gust speed. The group is always appended with units: KT (knots), MPS (meters per second) or KMH (km/h). If the wind direction will be variable, ddd is encoded as VRB. It is also permissible to encode the group as dddVddd to indicate a range of wind directions exceeding 60 degrees.

Prevailing visibility (VV). A number that may be whole or a fraction. Always ends with SM (statute miles) or nothing (meters).

Weather (ww). A two-letter standard abbreviation for any weather that will occur, with appropriate modifiers. The term CAVOK may be used if all clouds are above 5000 ft , visibility is above 10 km, and no significant precipitation is occurring; the United States does not use CAVOK.

Cloud condition (CCCHHH). Assigned for each cloud layer and may repeat. Consists of cloud cover (CCC) and height in hundreds of feet (HHH). Cloud cover may be clear, few (FEW, 1 to 2 eighths coverage), scattered (SCT, 3 to 4 eighths), broken (BKN, 5 to 7 eighths), or overcast (OVC). When the sky is obscured, CCC will be encoded as VV for vertical visibility and the HHH value will indicate the visibility into the obscuration.

Temperature/dewpoint (tt/dd). The temperature (tt) and dewpoint (dd) in whole degrees Celsius. If any value is negative, it is preceded by an "M".

Pressure (P). If this value starts with "A", it indicates altimeter setting with the value in hundreds of inches. If the value starts with "Q" it indicates sea-level pressure with the value in whole millibars.

Remarks (RMK). If the word "RMK" appears, it indicates that supplementary information follows.

```
2003/07/24 08:00
LPMT 240800Z 34009KT 9999 SCT018 21/17 Q1021 BLU

2003/07/24 08:00
LPOV 240800Z 33002KT 8000 BKN020 BKN080 19/16 Q1022 WHT

2003/07/24 08:00
LPST 240800Z 35009KT 9999 FEW008 BKN013 20/15 Q1023 GRN/WHT

2003/07/24 08:20
ESNU 240820Z 31006KT CAVOK 21/10 Q1012

2003/07/24 08:00
OIIP 240800Z 15005MPS 6000 SCT200 31/12 Q1007

2003/07/24 07:50
LTCG 240750Z 27013KT 9999 SCT035TCU BKN100 26/21 Q1010 NOSIG RMK RWY29
29014KT

2003/07/24 08:10
MYEG 240810Z AUTO 10007KT 10SM OVC001 27/23 A3004 RMK AO2 LTG DSNT NW
```

websites

weather.noaa.gov/weather/metar.shtml
adds.aviationweather.gov/metars
www.activitae.com/airbase/getMetar3.htm
www.uswx.com/us/stn

METAR remarks (U.S.)

A01 or A02. Automated station

SLPppp. Sea-level pressure, where ppp is the tens, units, and tenths value in millibars.

Tatttbdddd. Exact temperature (tttt) and dewpoint (dddd) in tens of degrees Celsius. The elements a and b are sign flags: when it is "1" the value that follows it is negative.

1xxxx. Six-hour max temperature (xxxx) in tens of degrees Celsius.

2nnnn. Six-hour minimum temperature (nnnn) in tens of degrees Celsius.

4/sss. Snow depth (sss)in whole inches.

4axxxbnnn. Twenty-four hour maximum (xxx) and minimum (nnn) temperature in tens of degrees Celsius. The elements a and b are sign flags: when it is "1" the value that follows it is negative.

5tppp. Pressure tendency (t) with 0-3 risen, 4 steady, and 5-8 fallen, and change (ppp) in tens of millibars.

6pppp. Six-hour precipitation (pppp) in hundreds of inches.

7pppp. Twenty-four hour precipitation (pppp) in hundreds of inches.

8lmh. Cloud type codes.

PCPN pppp or **P pppp**. One-hour precipitation (pppp) in hundreds of inches.

TAF forecast

The Terminal Aerodrome Forecast (TAF) is the worldwide standard for encoding standardized forecasts for any airport. It is based on the METAR observation format. For many decades the United States used domestic FT (terminal forecast) style, an extension of their SAO airways observation format. Both the FT and SAO formats were discontinued after a 1993-1995 transition period, and are now historical relics.

A TAF forecast can be particularly useful to meteorologists to ascertain what is expected at another location. The general format is as follows:

CCCC ddhhmm DDHHEE dddff VV ww CCCHHH

Station location (CCCC). This represents the four-letter ICAO identifier where the forecasted weather will occur.

Issuance time (ddhhmm). The UTC time the forecast was issued. The calendar day (dd); hour (hh); and minute (mm). A "Z" may be suffixed to the end as a reminder it is Zulu (UTC) time.

Forecast period (DDHHEE). The UTC time of the forecast period, with the starting day (DD) and hour (HH), and the ending hour (EE) (usually on the next day).

Wind (dddff). The wind direction in degrees relative to true north (ddd) and speed (ff). If winds are gusting, the group takes the form dddffGgg, where gg is the gust speed. The group is always appended with units: KT (knots), MPS (meters per second) or KMH (km/h). If the wind direction will be variable, ddd is encoded as VRB. It is also permissible to encode the group as dddVddd to indicate a range of wind directions exceeding 60 degrees.

Prevailing visibility (VV). A number that may be whole or a fraction. Always ends with SM (statute miles) or nothing (meters).

Weather (ww). A two-letter standard abbreviation for any weather that will occur, with appropriate modifiers. The term CAVOK may be used if all clouds are above 5000 ft , visibility is above 10 km, and no significant precipitation is occurring; the United States does not use CAVOK.

Cloud code group (CCCHHH). Assigned for each cloud layer and may repeat. Consists of cloud cover (CCC) and height in hundreds of feet (HHH). Cloud cover may be clear, few (FEW, 1 to 2 eighths coverage), scattered (SCT, 3 to 4 eighths), broken (BKN, 5 to 7 eighths), or overcast (OVC). When the sky is obscured, CCC will be encoded as VV for vertical visibility and the HHH value will indicate the visibility into the obscuration.

Wind shear group (WShhh/dddff). Sometimes, particularly in the United States, a low-level wind shear alert will be encoded. The value hhh specifies the maximum height above the surface in hundreds of feet, and ddd and ff specify the wind direction and speed above that height.

Transition identifier. These introduce new groups of weather conditions that will occur.

- FM hhmm indicates a significant change will take place at hour hh and minute mm.

- TEMPO hhee indicates a temporary condition lasting a total of less than half the time period will occur between hour hh and hour ee.

- BECMG hhee indicates a transition period that will begin at hour hh and end at hour ee, and after this time the new condition will become predominant.

- PROBpp hhee indicates a temporary condition with a probability value. The probability in percent is pp, and the duration of the expected weather ranges from hour hh to hour ee. This is used primarily in the United States. Only 30 or 40 is used; if there is a higher probability, then TEMPO is used.

Other groups. The U.S. military tends to use two groups:

- QNHppppINS, where pppp is the lowest expected altimeter setting in hundreds of inches. This is used by U.S. military stations.

- Ttt/hhZ is a maximum and minimum temperature group for the forecast period (two groups are used), where tt is the temperature and hh is the hour of occurrence. The group may also appear as TNtt/hhZ TXtt/hhZ.

```
2003/07/23 23:38
KJFK 232338Z 240024 18012G22KT 6SM BR SCT025 BKN050
      TEMPO 0004 4SM -SHRA BR OVC035
    FM0400 19012KT 4SM BR VCSH SCT020 BKN040
      TEMPO 0408 2SM SHRA BR OVC010
    FM0800 20009KT 3SM BR VCSH OVC015
      TEMPO 0812 1 1/2SM BR OVC008
    FM1200 20010KT 5SM BR BKN025
    FM1800 20012KT P6SM VCTS SCT025CB BKN050

2003/07/24 03:00
HTZA 240300Z 240606 17010KT 9999 SCT018 PROB 30 TEMPO 0608
      9000 -SHRA BKN016 SCT080 BECMG 1114 09015G20KT 9999 SCT020
    BECMG 1623 09010KT FEW018 BECMG 0005 13006KT SCT015

2003/07/24 04:00
HEMM 240400Z 240606 VRB03KT 6000 SKC FM0700 34014KT 9999 FEW020

2003/07/24 03:00
VECC 240300Z 240606 12008KT 4000 HZ SCT015 BKN090
      BECMG 1112 12006KT 3000 HZ
      BECMG 1618 VRB03KT 2000 HZ
      BECMG 0203 12005KT 3000 HZ
      BECMG 0506 12008KT 4000 HZ
      TEMPO 0606 1500 TSRA/RA SCT008 FEW025CB OVC080

2003/07/24 04:12
BIRK 240412Z 240606 22005KT 9999 SCT020 BKN040 BECMG 1518
          33008KT PROB40 TEMPO 1521 -SHRA BKN015 BECMG 2124
          33015KT 9000 -RA BKN015 OVC030

2003/07/24 04:00
UAAA 240400Z 240606 26005MPS 9999 BKN050CB BKN100
      TEMPO 0609 TS
      TEMPO 0913 VRB12MPS TSSHRA SQ
      TEMPO 1318 TSRA
      TEMPO 1822 SHRA
```

TEMP radiosonde data

Nearly all radiosonde data is transmitted in the TEMP format prescribed by the WMO in Publication 306, Section FM-35. It is broken up into three major blocks, TTAA (significant level), TTBB (mandatory level), and PPBB (winds aloft) data. Other blocks such as TTCC and TTDD pertain to data in the stratosphere and is not generally used by forecasters.

Mandatory level block (TTAA)

This block shows wind, temperature, and dewpoint at predesignated levels, such as 200 and 500 mb.

The block usually begins with:

TTAA ddhhi ccccc

The TTAA is a flag that shows this is the mandatory level block. Then follows the calendar day (dd) and hour (hh). A value of 50 is added to the day if the wind units are in knots, otherwise the wind units are in m/s. The highest wind data (i) is a coded figure which is roughly in hundreds of millibars. Finally the WMO station identifier (ccccc) is indicated.

Following this is a series of repeating blocks in the format:

pphhh TTTDD dddff

The level (pp) is expressed in tens of millibars, e.g. "85" indicates 850 mb. The exception is "99", which always is the first block and indicates the ground, and "92", which is 925 mb, and "88" and "77" are special use (see below). Following this is the height (hhh) in different expressions of meters, except for level "99" (ground) in which hhh is the surface pressure in tens, units, and tenths of a millibar. The hhh expression is in whole meters from 1000 to 700 mb (it is 1hhh meters at 850 mb and 2hhh or 3hhh meters at 700 mb, whichever brings it closer to 3000 m). From 500 to 400 mb hhh is expressed in decameters. From 300 to 100 mb hhh is 1hhh decameters. Following this is the temperature block TTTDD, with temperature (TTT) in tens of degrees Celsius and dewpoint depression (DD) in units and tenths of degrees Celsius if at or below "50" and in whole degrees Celsius if above

"50" (subtract 50 before using). Finally the wind is presented as direction (ddd) relative to true north and speed (ff). Direction always ends with "0" or "5", and "1" is added to it for each hundred units of wind speed (e.g. a dddff of 25604 indicates a wind from 255° at a speed of 104).

Tropopause information is encoded as 88ppp TTTDD dddff which indicates conditions at the tropopause: most importantly its pressure. Maximum winds are encoded as 77ppp dddff.

Significant level block (TTBB)

The significant level block is designed to show temperature and dewpoint only at levels bounded by strong changes. The header and format is much the same as the mandatory level (TTAA) block, except that the repeating data block is in the format:

nnppp TTTDD

where nn occurs in a repeating sequence (00 for the surface, followed by 11, 22, 33, 44, 55, 66, 77, 88, 99, 11, 22, etc). The rest of the block is identical to the TTAA block with the omission of wind data, and no tropopause or maximum wind data.

Winds aloft block (PPBB)

This block contains wind data only, and it is graduated in feet rather than millibars. Again, the header and format are similar to TTAA and TTBB format, except that repeating data is in the format:

9habc aaaAA bbbBB cccCC

The "9" is a marker that makes it easy to pick out the group elements. The rest of the group 9habc indicates heights of the block, followed by wind data aaaAA bbbBB cccCC (encoded the same way as in the TTAA/TTBB sections) at three levels. The ten-thousands place for height is indicated by h and the thousands place for each of the three groups by a, b, and c. The height value ha000 ft applies to wind group aaaAA, hb000 ft applies to wind group bbbBB, and hc000 ft applies to wind group cccCC. Not all three groups need to be encoded; if one or two are omitted, a, b, or c will contain a solidus.

```
520
USUS50 KWBC 241200 RRC

TTAA 74121 72318 99944 13606 26005 00145 ///// ///// 92814 14421
28513 85525 12657 30015 70126 02256 25526 50577 13563 24026 40743
23777 23056 30948 36173 22089 25072 44569 22101 20218 55167 21602
15401 55770 22556 10657 60771 21021 88183 56767 22083 77262 22105
41606 51515 10164 00004 10194 30015 27018=

647
UMUS41 KRNK 241217
SGLRNK

72318 TTBB   74120 72318 00944 13606 11931 14822 22877 11624
33869 13057 44850 12657 55700 02256 66631 01763 77620 02760
88578 05563 99509 12561 11470 17165 22430 19981 33196 55966
44112 56772 55100 60771 31313 45202 81106 41414 50961=

PPBB   74120 72318 90034 26005 30518 31517 90678 28515 28016
25517 909// 25522 91124 25523 25020 23519 916// 24023 92058
24525 23059 22582 93045 22587 22105 22102 9404/ 21602 22068
95024 22536 23528 21521=

445
USRE01 FMEE 241200

TTAA 74111 61976 99017 26260 12019 00157 22857 12021 92830 17025
12528 85548 13014 13032 70166 11091 14020 50590 04384 11515 40761
16179 10512 30972 30974 10509 25098 415// 04504 20246 537// 00514
15424 677// 02065 10662 807// 24508 88999 77149 02066 31313 47408
81059=

519
UKRE01 FMEE 241200

TTBB 74118 61976 00017 26260 11009 23458 22968 20034 33947 18229
44788 09205 55774 12459 66758 12091 77725 11091 88672 11291 99588
03086 11548 00885 22443 11380 33339 23776 44257 40171 55164 643//
66147 685// 77137 669// 88102 799// 21212 00017 12019 11867 12534
22803 14028 33766 11523 44702 14520 55641 12508 66555 13517 77443
09013 88328 13011 99216 00000 11149 02066 22142 01556 33/// /////
44118 30006 55104 18508 31313 47408 81059 41414 48501 51515 92830
17025 12528 77318 12258 12024 60439 04286 13012=
```

APPENDIX

Appendix 1A. Surface Plot Schematic

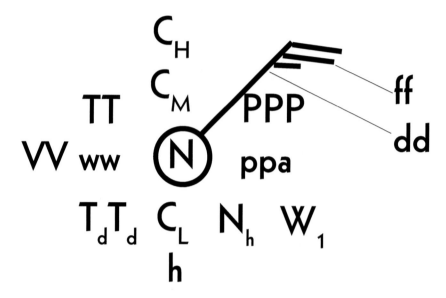

Presented here is the universally-adopted surface plot form.

TT — Temperature in degrees Celsius or Fahrenheit. Usually in whole degrees but may be expressed in tenths.

VV — Visibility in statute miles or meters. Mile values will appear as either whole numbers or fractional numbers. Meter values frequently appear as four digits, e.g. 0700.

ww — Symbol for weather type (see Table XXXX).

T_dT_d — Dewpoint temperature in degrees Celsius or Fahrenheit. Usually in whole degrees but may be expressed in tenths.

C_H — High cloud symbol.

C_M — Middle cloud symbol.

N — Total amount of cloud cover in oktas (eighths). The amount of the circle filled in is proportional to the amount of cloud cover. ○=Clear; ◔=1 okta; ◑=2 oktas; ◕=3 oktas; ◐=4 oktas; ◕=5 oktas; ◕=6 oktas; ◕=7 oktas; ●=8 oktas; ⊗=Sky obscured. When an automated station produced the observation, the symbol will be plotted as a square instead of a circle. Coloring may optionally be used: blue indicates MVFR flying conditions (ceiling 1000-3000 and/or visibility 3-5 sm); and red for IFR flying conditions (ceiling less than 1000 ft and/or visibility less than 3 miles).

C_L — Low cloud symbol.

h — Height of lowest low cloud layer, or if not present, lowest middle cloud layer. This is a coded single-digit value. 0=0-50 m; 1=50-100 m; 2=100-200 m; 3=200-300 m; 4=300-600m; 5=600-1000 m; 6=1000-1500 m; 7=1500-200 m; 8=2000-2500 m; 9=2500+ m; /=unknown.

PPP — Pressure in tens, units, and tenths of a millibar. Sometimes shows units, tenths, and hundredths of an inch of mercury.

pp — 3-hour pressure change in units and tenths of a millibar.

a — A two-segmented line representing pressure change during the past three-hours.

N_h — Amount of lowest low cloud layer, or if not present, lowest middle cloud layer. Expressed in oktas (eighths); if 9 the sky is obscured.

W_1 — Symbol for type of recent weather (See Table XXXX).

dd — Wind direction. Shaft points into the wind.

ff — Wind speed. A thick flag represents 50 kt, a long barb represents 10 kt, and each short barb represents 5 kt. The example shows 25 kt. If the wind is calm, the shaft is omitted and a circle is drawn around the station plot.

Appendix 1B. Upper Air Plot Schematic

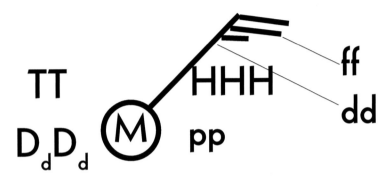

Presented here is the universally-adopted upper air plot form.

TT — Temperature in degrees Celsius. Usually in whole degrees but may be expressed in tenths.

D_dD_d — Dewpoint depression in Celsius degrees. Usually in whole degrees but may be expressed in tenths.

M — The plot circle is filled whenever the dewpoint depression is 5 Celsius degrees or less. This signifies the possible presence of cloud material and perhaps the threat for icing. The element will appear as a square when the observation was made by an aircraft (such as ACARS data) or a dropsonde. It will appear as an asterisk when the observation was satellite-based.

HHH — Geopotential height. It is either thousands, hundreds, and tens of meters or hundreds, tens, and units of meters according to the level involved. See the table at right.

pp — 12-hour height change in meters.

dd — Wind direction. Shaft points into the wind.

ff — Wind speed. A thick flag represents 50 kt, a long barb represents 10 kt, and each short barb represents 5 kt. The example shows 25 kt. If the wind is calm, the shaft is omitted and a circle is drawn around the station plot.

HEIGHT DECODING RULES

1000 mb: Geopotential height is a whole number in hundreds, tens, and units of meters. If the figure exceeds 500, subtract it from 500 to get the correct negative height.

925 mb: Geopotential height is a whole number in hundreds, tens, and units of meters.

850 mb: Geopotential height is in hundreds, tens, and units of meters. The thousands place is always "1".

700 mb: Geopotential height is in hundreds, tens, and units of meters. The thousands place is always "2" or "3", whichever brings the entire figure closest to 3,000.

500 mb, 400 mb: Geopotential height is in thousands, hundreds, and tens of meters. In other words, it is an expression in whole decameters.

300 mb, 250 mb: Geopotential height is in thousands, hundreds, and tens of meters. The ten-thousands place is always "0" or "1", whichever brings the entire figure closest to 10,000.

200 mb, 150 mb, 100 mb: Geopotential height is in thousands, hundreds, and tens of meters. The ten-thousands place is "1".

Appendix 2A. ICAO Regions

An ICAO identifier tells much more than you might expect, even when you don't know where the station is. The first digit always indicates the geographic region of the station. The second digit, in some cases, indicates the country or sub-region of the station. (Source: ICAO Document 7910, Location Indicators)

Loc	Region	Examples	
Axxx	Antarctica, New Guinea, Solomon Islands	AYPY	Port Moresby, Papua New Guinea
Bxxx	Greenland and Iceland	BIRK	Reykjavik, Iceland
Cxxx	Canada	CYYZ	Toronto, Ontario
Dxxx	Northwest Africa	DNMM	Lagos, Nigeria
Exxx	Northern Europe	EHAM	Amsterdam, Netherlands
Fxxx	Southern and Central Africa	FACT	Capetown, South Africa
Gxxx	West Africa & East Atlantic	GMTT	Tangier, Morocco
Hxxx	East Africa	HECA	Cairo, Egypt
Kxxx	United States	KJFK	New York City, New York
Lxxx	Southern Europe	LIRF	Rome, Italy
Mxxx	Central America and West Caribbean	MMMX	Mexico City, Mexico
Nxxx	South Pacific	NZAA	Auckland, New Zealand
Oxxx	Middle East	OIII	Teheran, Iran
Pxxx	North Pacific, Alaska, and Hawaii	PANC	Anchorage, Alaska
Rxxx	Western Pacific	RJAA	Tokyo, Japan
Sxxx	South America	SAEZ	Buenos Aires, Argentina
Txxx	Atlantic and East Caribbean	TJSJ	San Juan, Puerto Rico
Uxxx	Former Soviet Republics	UUEE	Moscow
Vxxx	India and Indochina	VIDP	Delhi, India
Wxxx	Indonesia, Malaysia, and Singapore	WSSS	Singapore
Yxxx	Australia	YSSY	Sydney, Australia
Zxxx	China, Mongolia, and North Korea	ZBAA	Beijing, China

Appendix 2B. WMO Regions

WMO station numbers are used primarily in SYNOP surface reports and in TEMP upper air observations. As with ICAO identifers, a WMO identifier can reveal some information about its location, as the first and second digits relate the station to a geographic area. (Source: WMO Pub. 9, Vol A - Observing Stations)

Loc	Region	Examples
0xxxx	Northwest Europe	07150 - Paris, France
1xxxx	Southeast Europe	16240 - Rome, Italy
2xxxx	Northern former USSR	27515 - Moscow, Russia
3xxxx	Southern former USSR	38457 - Tashkent, Uzbekistan
4xxxx	Middle East and Pacific Rim	47662 - Tokyo, Japan
5xxxx	China	54511 - Beijing, China
6xxxx	Africa	61641 - Dakar, Senegal
7xxxx	North America	72530 - Chicago, United States
8xxxx	South America	83378 - Brasilia, Brazil
9xxxx	Australasia	94767 - Sydney, Australia

Appendix 3. Descriptors

These are the standardized weather and obscuration to vision t ypes used in METAR and TAF forecasts. Only the SYNOP code format, containing up to 99 weather types, goes into further detail. The basic construction is in the order **intensity-proximity-precipitation-(space)-obscuration-miscellaneous.** Therefore freezing rain with fog is encoded as FZRA FG. All precipitation is assumed to be moderate unless a different intensity modifier (+ or -) is used. (Source: WMO Pub 306, Manual on Codes: Code Table 4678)

Abbv	Meaning	Type of item
-	Light	Intensity
+	Heavy	Intensity
BC	Patches	Descriptor
BL	Blowing	Descriptor
BR	Mist	Obscuration
DR	Drifting	Descriptor
DS	Dust storm	Miscellaneous
DU	Widespread dust	Obscuration
DZ	Drizzle	Precipitation
FC	Funnel cloud	Miscellaneous
FG	Fog	Obscuration
FU	Smoke	Obscuration
FZ	Freezing	Descriptor
GR	Hail	Precipitation
GS	Small hail	Precipitation
HZ	Haze	Obscuration
IC	Ice crystals	Precipitation

Abbv	Meaning	Type of item
MI	Shallow	Descriptor
PE	Ice pellets (sleet)	*DISCONTINUED*
PL	Ice pellets (sleet)	Precipitation
PO	Dust devils	Miscellaneous
PR	Partial	Descriptor
PY	Spray	Obscuration
RA	Rain	Precipitation
SA	Sand	Obscuration
SG	Snow Grains	Precipitation
SH	Shower	Descriptor
SN	Snow	Precipitation
SQ	Wind squalls	Miscellaneous
SS	Sandstorm	Miscellaneous
TS	Thunder	Descriptor
UP	Unknown precip	Precipitation
VA	Volcanic ash	Obscuration
VC	In vicinity	Proximity

Appendix 4. Present Weather

These two-digit numerical codes indicate the type of present weather that exists. They are used in SYNOP reports. The format is laid down in WMO Pub. 306, "Manual on Codes", table 4677.

Code	Sym	Meaning
00		Clear skies
01		Clouds dissolving
02		State of the sky unchanged
03		Clouds developing
04		Smoke
05	∞	Haze
06	S	Widespread dust not raised by wind
07	$	Dust or sand raised by wind
08	⑧	Dust devils
09	(S)	Duststorm or sandstorm not at station
10	=	Mist
11	⁼⁼	Patches of shallow fog
12	⁼⁼	Continuous shallow fog
13	<	Lightning visible, no thunder heard
14	⏺	Virga
15	)•(	Distant precipitation
16	(•)	Nearby precipitation
17	⟨R⟩	Thunderstorm with no precipitation
18	V	Wind squall
19	)	Funnel cloud, waterspout, or tornado
20	)	Drizzle during past hour
21	•]	Rain during past hour
22	*]	Snow during past hour
23	*]	Rain and snow during past hour
24	~]	Freezing rain during past hour
25	⏗]	Rain showers during past hour
26	⏗]	Snow showers during past hour
27	⏗]	Hail showers during past hour
28	≡]	Fog during past hour
29	⟨R⟩]	Thunderstorm during past hour
30	S⏗	Slight-moderate duststorm, decreasing
31	S	Slight-moderate duststorm, steady
32	⎸S	Slight-moderate duststorm, increasing
33	S⏗	Severe duststorm, decreasing
34	S	Severe duststorm, steady
35	⎸S	Severe duststorm, increasing
36	+	Slight-moderate drifting snow
37	+	Heavy drifting snow
38	+	Slight-moderate blowing snow
39	+	Heavy blowing snow
40	⊜	Fog at a distance
41	≕	Patches of fog
42	≡⎸	Fog, sky visible, thinning
43	≡	Fog, sky not visible, thinning
44	≡	Fog, sky visible, no change
45	≡	Fog, sky not visible, no change
46	⎸≡	Fog, sky visible, becoming thicker
47	⎸≡	Fog, sky not visible, becoming thicker
48	⩝	Fog, depositing rime, sky visible
49	⩝	Fog, depositing rime, sky not visible
Code	Sym	Meaning
50	,	Drizzle, light, intermittent
51	,,	Drizzle, light, continuous
52	⁏	Drizzle, moderate, intermittent
53	⋰	Drizzle, moderate, continuous
54	⦙	Drizzle, heavy, intermittent
55	⸪	Drizzle, heavy, continuous
56	⸱∿	Freezing drizzle, light
57	∿	Freezing drizzle, moderate or heavy
58	⁏	Drizzle and rain, light
59	⁏	Drizzle and rain, moderate or heavy
60	•	Rain, light, intermittent
61	••	Rain, light, continuous
62	:	Rain, moderate, intermittent
63	∴	Rain, moderate, continuous
64	⦙	Rain, heavy, intermittent
65	⸪	Rain, heavy, continuous
66	∿	Freezing rain, light
67	∿	Freezing rain, moderate or heavy
68	⁞	Rain and snow, light
69	⁞	Rain and snow, moderate or heavy
70	*	Snow, light, intermittent
71	**	Snow, light, continuous
72	⁂	Snow, moderate, intermittent
73	⁂	Snow, moderate, continuous
74	⦙	Snow, heavy, intermittent
75	⸪	Snow, heavy, continuous
76	—	Diamond dust (ice crystals)
77	⇢	Snow grains
78	⇠	Snow crystals
79	△	Ice pellets
80	▽	Rain showers, light
81	▽	Rain showers, moderate to heavy
82	▽	Rain showers, violent
83	▽	Snow and rain showers, light
84	▽	Snow and rain showers, moderate to heavy
85	▽	Snow showers, light
86	▽	Snow showers, moderate to heavy
87	▽	Snow and ice pellet showers, light
88	▽	Snow and ice pellet showers, mod. to heavy
89	▽	Hail showers, light
90	▽	Hail showers, moderate to heavy
91	⟨R⟩	Recent thunderstorm, light rain
92	⟨R⟩	Recent thunderstorm, mod. to heavy rain
93	⟨R⟩	Recent thunderstorm, light snow or mix
94	⟨R⟩	Recent thunderstorm, mod-heavy snow/mix
95	R	Thunderstorm, light to moderate
96	R	Thunderstorm, light to moderate w/ hail
97	R	Thunderstorm, heavy
98	R	Thunderstorm, heavy, with duststorm
99	R	Thunderstorm, heavy, with hail

Appendix 5. Cloud code groups

These are the code forms used to represent low, middle, and high clouds. They are commonly encoded in both SYNOP and METAR reports. (Source: WMO Pub 306, Manual on Codes: Code Table 509, 513, and 515)

Low cloud types
1   CUMULUS, fair weather, no vertical development
2   CUMULUS, moderate vertical development
3   CUMULONIMBUS, no anvil
4   STRATOCUMULUS formed by spreading of cumulus
5   STRATOCUMULUS
6   STRATUS, of fair weather
7   STRATUS, of bad weather (scud)
8   CUMULUS AND STRATOCUMULUS with bases at different levels
9   CUMULONIMBUS with anvil cloud
0   No low clouds
/   Low clouds not visible due to darkness or obscuration

Middle cloud types
1   ALTOSTRATUS, mostly transparent
2   ALTOSTRATUS, opaque, or NIMBOSTRATUS
3   ALTOCUMULUS, mostly transparent
4   ALTOCUMULUS, patches
5   ALTOCUMULUS, invading the sky
6   ALTOCUMULUS, formed by spreading of cumulus
7   ALTOCUMULUS, at different layers
8   ALTOCUMULUS, castellanus (cumuliform)
9   ALTOCUMULUS, of a chaotic sky at random levels
0   No middle clouds
/   Middle clouds not visible due to darkness or obscuration

High cloud types
1   CIRRUS, fibrous
2   CIRRUS, in dense patches
3   CIRRUS, from cumulonimbus anvil
4   CIRRUS, progressively invading the sky
5   CIRRUS OR CIRROSTRATUS, invading sky, less than 45 deg above horizon
6   CIRRUS OR CIRROSTRATUS, invading sky, more than 45 deg above horizon
7   CIRROSTRATUS, covering the entire sky
8   CIRROSTRATUS, not covering the entire sky, not invading
9   CIRROCUMULUS
0   No high clouds
/   High clouds not visible due to darkness or obscuration

Appendix 6. Isopleths

What is a line called when it represents a certain quantity? This table will explain the technical name. (Source: AMS Glossary of Meteorology <amsglossary.allenpress.com/glossary>).

A line of equal . . .	Term
Temperature	isotherm
Potential temperature (theta)	isentrope
Dewpoint	isodrosotherm
Humidity	isohume
Wind speed	isotach, isovel
Wind direction	isogon
Shear	isoshear
Pressure	isobar
Density	isopycnal, isopycnic
Height	isoheight, contour, isohypse
Cloud cover	isoneph
Time	isochrone
Thunderstorm phase	isobront
Thunderstorm frequency or intensity	isoceraunic
Radar Doppler velocity	isodop
Precipitation	isohyet
Seasonal precipitation	isomer
Snowfall or snow depth	isonival, isochion
Sunlight	isohel
Aurora frequency	isochasm
Radar echo intensity	isoecho

Appendix 7. Chart Analysis Symbology

Shown here are standard markings used by Air Force Global Weather Central during the 1950s and 1960s, as published by Col. Robert C. Miller in Notes on Analysis and Severe-Storm Forecasting Procedures of the Air Force Global Weather Central (1972). Miller was a driving force in severe weather forecasting during the 1950s and 1960s and established many of the techniques used in Air Force forecasting during that era. He also created one of the few sets of meteorological symbology ever developed. While some styles have been adopted, many have become technically obsolete, fallen into disuse, or substituted with generic styles. Regardless of the state of modern-day techniques, it can be said that there has been no comparable set of conventions released since Miller's 1972 paper, and they serve as a fascinating reference.

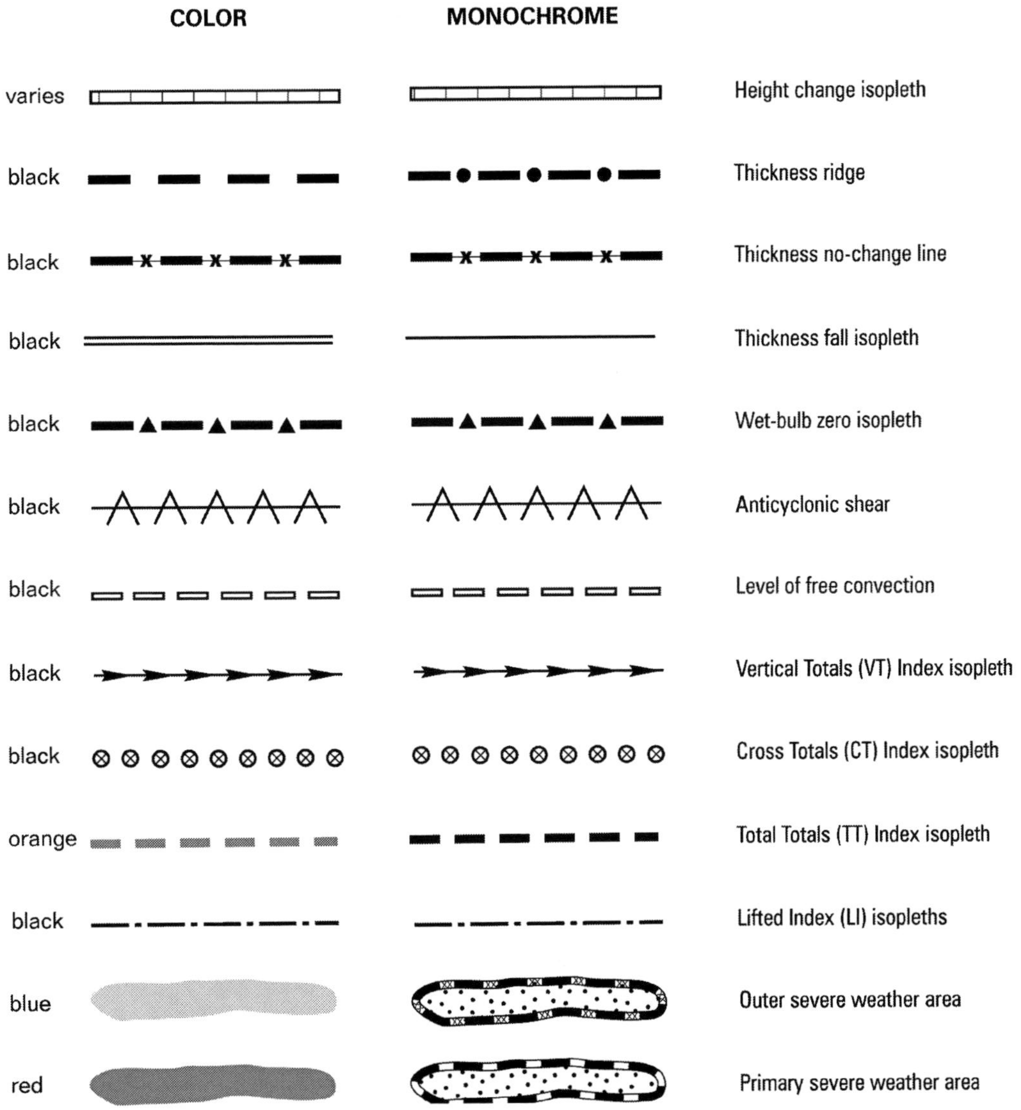

COLOR	MONOCHROME	
varies		Height change isopleth
black		Thickness ridge
black		Thickness no-change line
black		Thickness fall isopleth
black		Wet-bulb zero isopleth
black		Anticyclonic shear
black		Level of free convection
black		Vertical Totals (VT) Index isopleth
black		Cross Totals (CT) Index isopleth
orange		Total Totals (TT) Index isopleth
black		Lifted Index (LI) isopleths
blue		Outer severe weather area
red		Primary severe weather area

COLOR	MONOCHROME	
green		850 mb isodrosotherm/isohume
green		850 mb moisture axis
green 35 KT	35 KT	850 mb jet axis
red		850 mb dryline / dry prod / dry intrusion edge
red		850 mb temperature ridge
red		850 mb axis of cold air advection
red 25 KT 35 KT	25 KT 35 KT	850 mb shear
brown		700 dry intrusion edge
brown		700 mb moisture
brown — x — x — x —	— x — x — x —	700 mb 12-hr no change (T or hgt)
brown ● x ● x ● x ● x ● x ●	● x ● x ● x ● x ●	700 mb temperature ridge
brown △ △ △ △ △	△ △ △ △ △	700 mb thermal trough
brown		700 mb convergence zone
brown		700 mb axis of cold air advection
brown 55 KT	55 KT	700 mb jet axis (dry)
brown		700 mb diffluence
brown		700 mb significant height falls
brown		700 mb significant temperature falls

COLOR		MONOCHROME	
blue	▬ ▬ ▬ ▬	○○○○○○○○○○○	500 mb isotherms
blue	▬ ▬ ▬ ▬	◇◇◇◇◇	500 mb critical isotherm
blue	△ △ △ △ △	△ △ △ △ △	500 mb thermal trough
blue	-16	-16	500 mb significant height falls
blue	-4	-4	500 mb significant temperature falls
green			500 mb moisture
yellow			500 mb PVA zone
blue	70 KT →	70 KT ⇒	500 mb jet axis
blue	∿	∿	500 mb shear
blue	⋀⋁⋀⋁	⋀⋁⋀⋁	500 mb diffluence
purple	90 KT →	90 KT ⇒	300-200 mb jet axis
purple	◆	◇	300-200 mb jet max
purple	⋀⋁⋀⋁	⋀⋁⋀⋁	300-200 mb diffluence
purple	∿	∿	300-200 mb shear

Appendix 8. Stability Indices

The long, hard road of understanding the thunderstorm is demonstrated by the plethora of stability indices. Many of them exist as simple rules of thumb, created in a time where calculators and slide rules precluded lengthy calculations for multiple forecast points. Since the 1990s, the explosion of computing power has made computations of integrated stability (CAPE) and shear a relatively trivial process. Therefore the stability parameters in use today are comprised mainly of CAPE, CINH, SRH, and EHI. However the complete set of indices are presented here for the reader's information. It must be remembered that these are all simplifications of very complex processes, and in no way do they replace a meaningful understanding of the sounding. The equations used here are presented in as simple a format as possible. Terms used are T=temperature (deg C), T_d=dewpoint (deg C), D=dewpoint depression (C deg), FF=wind speed (kt); DD=wind direction (deg C). Other terms are explained where they occur.

■ Vertical Totals Index (VT)

The Vertical Totals Index is a measure of the lapse rate from about 5,000 to about 18,000 ft in the atmosphere. It makes no assumptions about parcel temperature. The VT Index was published by Robert Miller in 1967.

VTI = T(850) - T(500).

25 - Storms are unlikely

26 - Scattered thunderstorms

30 - Scattered thunderstorms, a few severe, isolated tornadoes

32 - Scattered to numerous thunderstorms, scattered to a few severe, a few tornadoes

34+ - Numerous thunderstorms, scattered severe storms, scattered tornadoes

■ Cross Totals Index (CT)

The Cross Totals Index relates the low-level moisture to the mid-level temperature, yielding an indirect estimate of lapse rate and convective instability. It was published by Robert Miller in 1967. The Surface Based Cross Totals Index (SCTI) is an alternative index that uses the surface instead of 850 mb.

CTI = T(850) - T(500).

<17 - Thunderstorms unlikely

18 to 19 - Isolated to few thunderstorms

20 to 21 - Scattered thunderstorms

22 to 23 - Scattered thunderstorms, isolated severe

24 to 25 - Scattered thunderstorms, few severe, isolated tornadoes

26 to 29 - Scattered to numerous thunderstorms, few to scattered severe, few tornadoes

30+ - Numerous thunderstorms, scattered severe, scattered tornadoes

■ Total Totals Index (TT)

The Total Totals Index attempts to integrate the lapse rate information of the Vertical Totals Index with the instability information in the Cross Totals Index. The index is sensitive to steep lapse rates, even if insufficient moisture is present. It was devised by Robert Miller in 1967.

TT = VT + CT

<43 - Thunderstorms unlikely

44-45 - Isolated or few thunderstorms

46-47 - Scattered thunderstorms

48-49 - Scattered thunderstorms, isolated severe

50-51 - Scattered heavy thunderstorms, few severe, isolated tornadoes

52-55 - Scattered to numerous heavy thunderstorms, few to scattered severe, few tornadoes

56+ - Numerous heavy thunderstorms, scattered severe, scattered tornadoes

■ K Index (KI)

This parameter is a sum of lapse rate, 850 mb moisture, and humidity at 700 mb. Humidity at 700 mb is a significant contribution, though this is rare with Great Plains storm events or any event that depends on a mid-level cap. The index was published by J. J. George in 1960.

KI = T(850) - T(500) + T_d(850) - D(700)

<15 - No thunderstorms (0%)

15-20 - Thunderstorms unlikely (<20%)

21-25 - Isolated thunderstorms (20-40%)

26-30 - Widely sct. thunderstorms (40-60%)

30-35 - Numerous thunderstorms (60-80%)

36-40 - Numerous thunderstorms (80-90%)

40+ - Definite thunderstorms (100%)

■ Showalter Stability Index (SSI)

The SSI lifts a parcel from 850 mb to 500 mb. It has the advantage of avoiding the problems inherent with shallow moisture situations, however this comes at the cost of ignoring boundary-layer characteristics. It does not work well in mountainous areas, and cannot be used when the 850 mb level is below ground level. The SSI was developed by Albert Showalter in 1947.

$$SSI = T_{ENVIR}(500) - T_{PARCEL}(500)$$
for a parcel lifted from 850 mb

>+3: No thunderstorms likely

+3 to +1: Showers probable, thunderstorms possible

0 to -3: Moderate indication of severe thunderstorms

-4 to -6: Strong indication of severe thunderstorms

<-6: Severe thunderstorms likely

■ Lifted Index (LI)

The Lifted Index was widely favored in 1980s before integrated stability measures became widespread. It lifts a parcel from the surface to 500 mb and compares the parcel temperature to the environmental temperature. The parcel's starting dewpoint should originate from an average mixing ratio of the lowest 100 mb and from the forecast afternoon temperature. The historical origin of the Lifted Index is attributed to Joseph Galway in 1956.

$$LI = T_{ENVIR}(500) - T_{PARCEL}(500)$$
Parcel lifted from surface

>2 - No significant activity

2 to 0: Showers/thunderstorms possible with other source of lift

0 to -2: Thunderstorms possible

-2 to -4: Thunderstorms probable, only a few severe

<-4: Severe thunderstorms possible

■ Modified Lifted Index (MLI)

The Modified Lifted Index is the same as the Lifted Index except that the parcel is lifted from the highest wet bulb temperature in the lowest 300 mb of the atmosphere. The index was developed by Charles Doswell to better forecast thunderstorms on the High Plains. It was presented as a possible local enhancement to the lifted index, but was adopted for use by the NWS and military forecasters.

$$MLI = T_{ENVIR}(500) - T_{PARCEL}(500)$$
parcel lifted from maximum wet bulb temperature in lowest 300 mb of atmosphere

Positive: No thunderstorms likely

0 to -2: Showers probable, thunderstorms possible

-3 to -5: Moderate indication of thunderstorms

<-6: Strong indication of severe thunderstorms

Another type of modified lifted index exists which raises a parcel to the -20 deg C isotherm instead of to the 500 mb level. The underlying concept is that a significant thunderstorm should have a cloud temperature of -20 deg C. The historical source of this index is not known.

■ Thompson Index (TI)

The Thompson Index is a combination of Lifted Index and K Index. It attempts to integrate elevated moisture into the index, using the 850 mb dewpoint and 700 mb humidity. Accordingly, it works best in tropical and mountainous locations. The historical origin of the index is not known.

$$TI = K - LI$$

<u>Over Rockies</u>

<20: Thunderstorms unlikely

20-29: Thunderstorms

30-34: Thunderstorms approaching severe

35+: Severe thunderstorms

<u>East of Rockies</u>

<25: Thunderstorms unlikely

25-34: Slight chance of thunderstorms

35-39: Few widely scattered thunderstorms approaching severe

>40: Severe thunderstorms

■ Severe Weather Threat Index (SWEAT)

The SWEAT index uses a complex set of parameters. It was one of the first indices developed specifically to assess tornado potential. Important parameters are 850 mb dewpoints, lapse rates and parcel instability, wind speed at 850 mb and 500 mb, and directional shear betwen 850 mb and 500 mb. The SWEAT index was published in 1972 by Robert Miller. The index uses the Total Totals Index, which must be computed first.

$$SWEAT = 12 \times Td(850) + 20 \times (TT-49) + 2 \times FF(850) + FF(500) + 125 \times (\sin[DD(500) - DD(850)] + 0.2)$$

<300: Non-severe thunderstorms

300-400: Isolated moderate to heavy thunderstorms

400-500: Severe thunderstorms and tornadoes probable

500-800: Severe thunderstorms and tornadoes likely

800+: Possibly no severe weather (sheared convection)

■ Convective Available Potential Energy (CAPE)

CAPE is currently the most widely used predictor for both thunderstorm potential and severe weather risk. It was defined in 1982 by Morris Weisman and Joseph Klemp. The form of the

$$CAPE = \left(\sum_{LFC}^{EL} \left[\frac{(T_{ap} - T_e)}{T_e} \vec{g} \right] \right) \Delta z$$

equatio is shown here:

which yields integrated instability in joules per kilogram.

<300: Mostly stable, little or no convection
300-1000: Marginally unstable; weak thunderstorm activity
1000-2500: Moderately unstable; possible severe thunderstorms
2500-3500: Very unstable; severe thunderstorms; possible tornadoes
3500+: Extremely unstable; severe thunderstorms; tornadoes likely

■ Convective Inhibition (CINH)

Convective Inhibition is calculated in the same manner as CAPE except for areas along the parcel lift where the parcel is colder than the surrounding air. In effect, it figures areas that are negatively buoyant. It was developed in 1984 by Frank Colby.

<0: No cap
0 to 20: Weak capping
21-50: Moderate capping
51-99: Strong capping
100+: Intense cap; storms not likely

■ Bulk Richardson Number (BRN)

The Bulk Richardson Number is a ratio between instability and 0-6 km vertical shear. It is a discriminator of storm type, not a predictor. High values indicate unstable and/or weakly sheared environments, while low values indicate weak instability and/or strong shear. It was defined in 1986 by Morris Weisman and Joseph Klemp.

BRN = CAPE / [0.5 * U²]
    where U is the difference in wind speed in meters per secnd between 0 and 6 km

<10: Severe weather unlikely
10-45: Associated with supercell development
>50: Weak multicell storms

■ BRN Shear

BRN shear is simply a measure of the vector difference in the winds through the vertical. The greater the BRN shear, the more likely that a thunderstorm downdraft and precipitation cascade will be separated from the updraft.

BRN Shear = 0.5 (U$_{AVG}$)²
    where UAVG is the vector difference between the 0 to 6 km AGL winds and the winds in the lowest 0.5 km of the atmosphere

25-50: Sometimes associated with tornadic storms
50-100: Associated with tornadic storms

■ Energy-Helicity Index (EHI)

EHI is a product of CAPE and 0-6 km shear, thus it is high when either parameter is high. It was developed in 1991 by John Hart and Josh Korotky.

EHI = (CAPE x SRH) / 160,000
1.0-2.0: Heightened threat of tornadoes
2.0-2.4: Tornadoes possible but unlikely strong
2.5-2.9: Tornadoes likely
3.0-3.9: Strong tornadoes possible
4.0+: Violent tornadoes possible

■ Storm Relative Helicity (SRH)

EHI is a product of CAPE and 0-6 km shear, thus it is high when either parameter is high. The SRH is not used widely because it has a high dependence on the correct storm motion vector and it is extremely sensitive to the wind field. The SRH was defined in 1990 by Robert Davies-Jones, Don Burgess, and Mike Foster.

SRH = w · (v - c) Δz
    where w = k x dV/dz from 0 to 6 km, v is the wind vector, and c is the storm vector

150-299: Weak tornado potential
300-449: Moderate tornado potential
450+: Strong tornado potential

■ KO Index

This index was developed by Swedish meteorologists and used heavily by the Deutsche Wetterdienst. It compares values of equivalent potential temperature at different levels. It was developed by T. Andersson, M. Andersson, and C. Jacobsson, and S. Nilsson.

KO = 0.5 x (θ_e(500) + θ_e(700)) - 0.5 x (θ_e(850) + θ_e(1000))

>6: No thunderstorms
2-6: Thunderstorms possible
<2: Severe thunderstorms possible

Boyden Index (BI)

This index, used in Europe, does not factor in moisture. It evaluates thickness and mid-level warmth. It was defined in 1963 by C. J. Boyden.

BI = Z(700) - Z(1000) - T(700) - 200

    where Z is height in dam

>95: Thunder possible

Bradbury Index (BRAD)

Also known as the Potential Wet-Bulb Index, this index is used in Europe. It is a measure of the potential instability between 850 and 500 mb. It was defined in 1977 by T. A. M. Bradbury.

$BRAD = \theta_w(500) - \theta_w(850)$

<3: Thunderstorms possible

Rackliff Index (RI)

This index, used primarily in Europe during the 1950s, is a simple comparision of the 900 mb wet bulb temperature with the 500 mb temperature. It is believed to have been developed by Peter Rackliff during the 1940s.

$RI = \theta_w(900) - T_{500}$

>30: Thunderstorms possible

Jefferson Index (JI)

A European stability index, the Jefferson Index was intended to be an improvement of the Rackliff Index. The change would make it less dependent on temperature. The version in use since the 1960s is a slight modification of G. J. Jefferson's 1963 definition.

$JI = 1.6 \times \theta_w(850) - T(500) - 0.5 \times (T(700) - T_d(700)) - 8$

>30: Thunderstorms possible

S-Index (S)

This European index is a mix of the K Index and Vertical Totals Index. It was designed to be an optimized vertion of the Total Totals Index. The S-Index was developed by the German Military Geophysical Office.

S = KI - (T(500) + A)

    where A is 0 if the VT is greater than 25, 2 if the VT is between 22 and 25, and 6 if the VT is less than 22.

<39: No thunderstorms

41-45: Thunderstorms possible

>46: Thunderstorms likely

Yonetani Index (YON)

This index was developed by Japanese meteorologist Tsuneharu Yonetani in 1979 to forecast thunderstorms on the Kanto Plain. It provides a measure of conditional instability and low level moisture.

$YON = 0.966\Gamma_L + 2.41(\Gamma_U - \Gamma_W) + 0.966\gamma - 15$

The final term is 16.5 instead of 15 if γ is less than or equal to 0.57. Γ is the layer lapse rate, with U representing 850-500 mb and L representing 900-850 mb, and W is the lapse rate at 850 mb. The term γ is the pressure weighted average of the relative humidity in the 900-850 mb layer, ranging from 0-1.

>0: Thunderstorms likely

Potential Instability Index (PII)

This index relates potential instability in the middle atmosphere with thickness. It was defined by A. J. Van Delden in 2001.

$PII = (\theta_e(700) - \theta_e(500)) / (Z(500) - Z(925))$

>0: Thunderstorms likely

Deep Convective Index (DCI)

This index is a combination of parcel theta-e at 850 mb and lifted index. This attempts to further improve the lifted index. It was defined by W. R. Barlow in 1993.

$DCI = T(850) + T_d(850) - LI$

10-20: Weak thunderstorms

20-30: Moderate thunderstorms

30+: Strong thunderstorms

Appendix 9. Miller's Severe Weather Parameters

In 1972, Col Robert C. Miller published a list of parameters used by the Air Force Global Weather Central to identify thunderstorm risk areas. It was a basis for techniques used within the National Weather Service and worldwide for decades. Though slightly outdated and not a true ingredients-based approach, it is presented here for its informational value.

Rank	Parameter	Weak	Moderate	Strong
1	500 mb Vorticity	Neutral or Negative Advection	Contours cross vort pattern by <30 deg	Contours cross at more than 30 deg
2	Lifted Index Total Totals	-2 <50	-3 to -5 50 to 55	-6 >55
3	Mid-Level Jet Mid-Level Shear	<35 kt <15 kt per 90 nm	35-50 kt 15-30 kt per 90 nm	>50 kt >30 kt per 90 nm
4	Upper-Level Jet Upper-Level Shear	<55 kt <15 kt per 90 nm	55 to 85 kt 15-30 kt per 90 nm	>85 kt >30 kt per 90 nm
5	Low-level jet	<25 kt	25-34 kt	>34 kt
6	Low-level moisture	<8 g/kg	8-12 g/kg	>12 g/kg
7	850 mb max temp. field	E of moist ridge	over moist ridge	W of moist ridge
8	700 mb height no-change line	Winds cross line <20 deg	Winds cross line 20-40 deg	Winds cross line >40 deg
9	700 mb dry air intrusion	Not available or weak winds	Winds from dry to moist intrude at an angle of 10 to 40 deg are at least 15 kt	Winds intrude at an angle of 40 deg and are at least 25 kt
10	12 hr surface pressure falls	<1 mb	1-5 mb	5 mb
11	500 mb height chg	<30 m	30-60 m	>60 m
12	Height of wet-bulb zero above surface	Above 11,000 ft Below 5,000 ft	9,000-11,000 ft or 5,000-7,000 ft	7,000-9,000 ft
13	Surface pressure over threat area	>1010 mb	1010-1005 mb	<1005 mb
14	Surface dewpoint	<55 deg F	55-64 deg F	>64 deg F

Appendix 10. Satellite Gallery

This section shows some of the major cloud forms as seen on typical high-resolution satellite imagery.

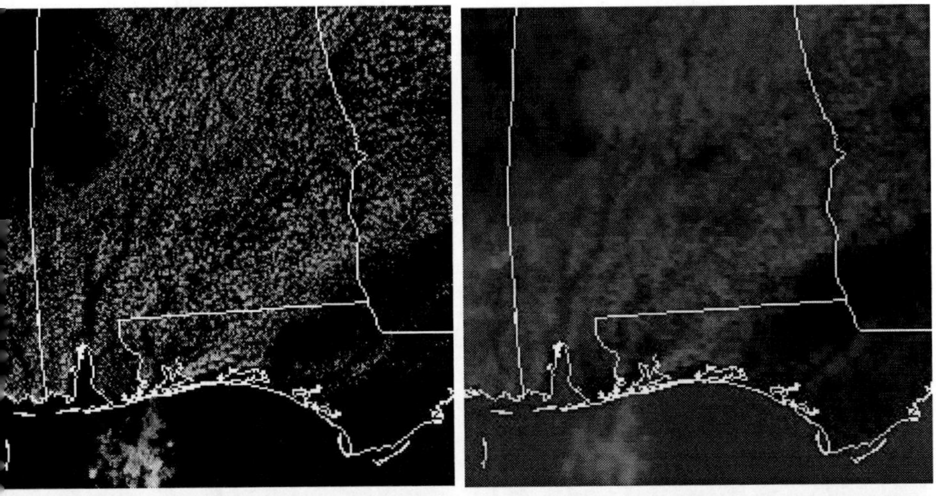

Cumulus usually appears in fields, as seen here, though it may prefer to develop along mountains. Cumulus fields have a random speckled appearance on visible imagery. The cumulus tops are not very cold, and on normal 4 km infrared imagery they tend to appear as a gray blur with speckled structure.

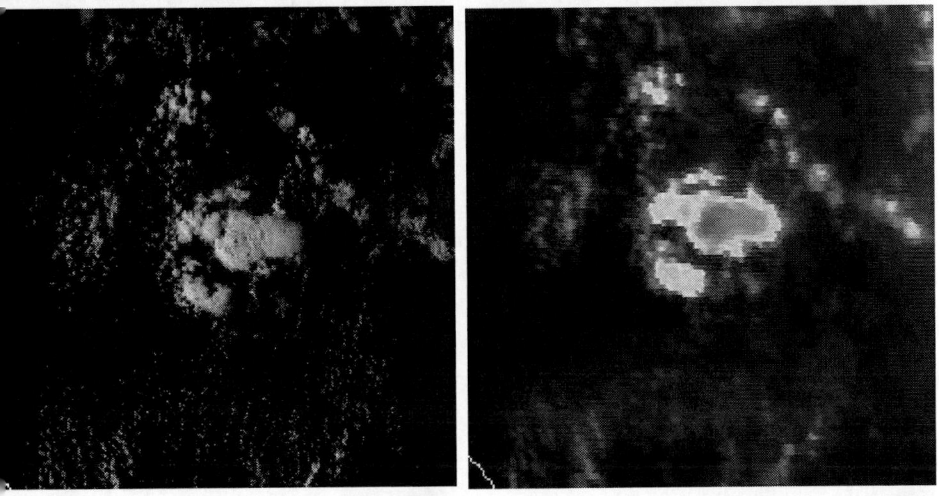

Isolated cumulonimbus is quite distinctive. Visible imagery shows very bright globular cloud masses which give way to a very bright circular or elongated anvil cloud. Infrared imagery shows small cloud elements with very cold temperatures; this expands over hours into a large, cold cirriform mass.

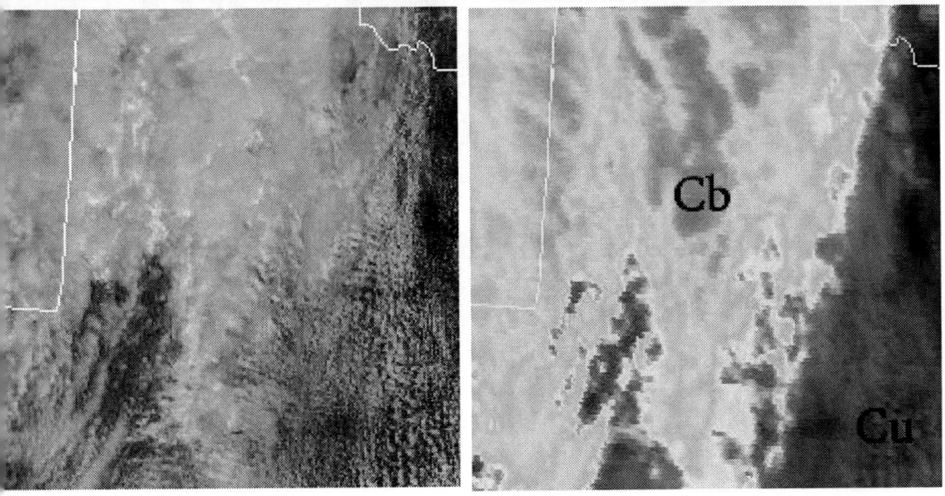

Embedded cumulonimbus is produced by significant synoptic-scale lift in marginal instability. Visible imagery shows multiple layers with mottled elements. Infrared shows organized spots of very cold tops. The majority of the cold cloud is actually cirriform anvil debris.

Stratocumulus appears as a bright white cloud with speckled edges indicative of cumulus. Infrared imagery shows a dull warm cloud, generally with soft edges. Some cirrus overlies this example, shown by the enhanced spots of yellow and green on the IR image.

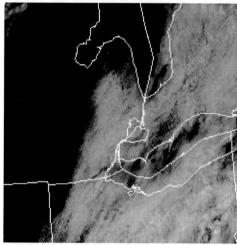
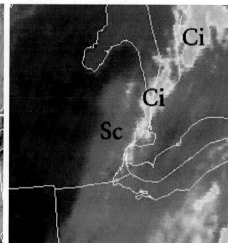

Stratus and fog looks on visible imagery like a featureless white cloud with edges that seem to be influenced by terrain; fog follows valley outlines. Infrared shows a very warm cloud that's dull, featureless gray; if warm lakes can be seen, then the cloud is probably fog and not stratus.

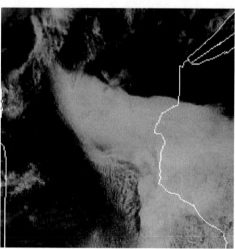
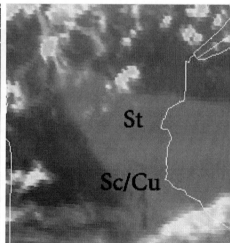

Nimbostratus, the classic rain cloud, does not occur by itself but is a amalgamation of very thick altostratus and cirrus layers. Visible imagery shows a large-scale overcast; infrared shows a moderately cold cloud with much the same. Numerous cold spots suggest more of a convective than stratiform situation.

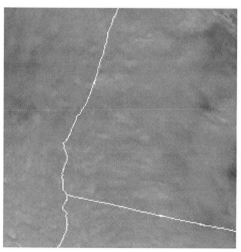
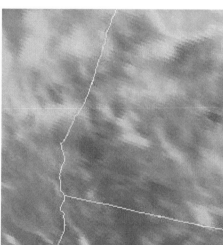

VISIBLE	INFRARED

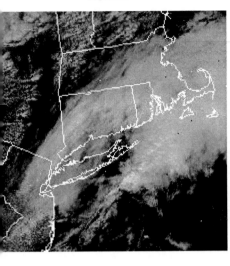 

Altostratus is a fibrous bright cloud on visible imagery, rarely with sharp edges. Infrared imagery shows a marginally cool, featureless cloud. It usually occurs in conjunction with other layers. The example here appears to have breaks in the altostratus, with stratocumulus underneath.

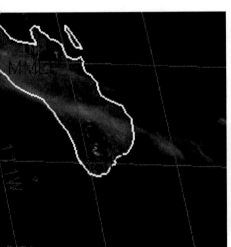 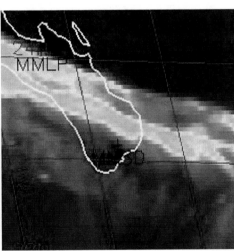

"Fair weather cirrus" has markedly different appearances on different channels. Visible imagery shows a veil-like layer with soft edges, which may have a patchy appearance. Infrared imagery shows a very cold cloud mass with soft edges and no concentrated cold spots that would be indicative of cumulonimbus underneath.

 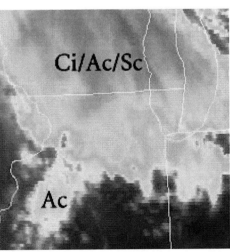

Embedded cirrus with other underlying layers appears bright on visible imagery but has the characteristic soft, fibrous appearance of cirrus. Infrared imagery shows a large, cold mass of cloud. This configuration is common with well-developed weather disturbances.

Glossary

This section provides a summary of acronyms that may be encountered in journals, case studies, and forecasting discussions.

ablation Depletion of snow and ice by melting and evaporation.

Ac Altocumulus (q.v.)

AC 1. Convective Outlook bulletin; from the Family of Services data stream header. 2. Altocumulus

ACARS Aircraft Communications and Reporting System. A block of data transmitted by aircraft that often contains weather data.

ACCAS Altocumulus Castellanus. Altocumulus which forms in a convectively unstable layer.

accessory cloud A cloud which is dependent on a larger cloud system for development.

ACSL Altocumulus standing lenticular (q.v.)

acre-foot The amount of water required to cover one acre to one foot of depth. This equals 326,851 gallons or 43,560 cubic feet.

adiabatic The change in temperature without a transfer of heat. It may be caused by compression or expansion.

advection Horizontal movement of air that causes changes in the physical properties of air at a specific location.

advection fog Fog that forms as warmer, moist air moves over a cold surface. The air is forced to condense as it loses heat to conduction.

advisory In the United States, a weather bulletin which is less serious than a warning.

AFD Area Forecast Discussion (q.v.)

AFGWC Air Force Global Weather Central, which became obsolete on 15 October 1997. It is now known as AFWA.

AFOS Automation of Field Operations and Services (discontinued). The backbone computer system of NWS offices; developed in 1976 and fielded in 1979; retired between 1996 and 1999.

AFWA Air Force Weather Agency (q.v.)

AGL Above Ground Level.

Air Force Weather Agency (AFWA) Weather component of the U.S. Air Force. Activated 15 October 1997, it is a combination of Air Weather Service Headquarters and the Air Force Global Weather Center.

albedo The portion of incoming radiation which is reflected by a surface.

air mass A body of air which contains relatively uniform properties of temperature and moisture.

AJ Arctic jet (q.v.)

algorithm A computer program designed to solve a specific problem. Often used in WSR-88D radars.

aliasing A process in which a radar return has a frequency too high to be analyzed within the given sampling interval but at a frequency less than the Nyquist interval.

ALSTG Altimeter setting (q.v.)

altimeter setting The pressure at which an altimeter must be set so that it reads the correct elevation.

altocumulus Mid-level clouds composed primarily of water or supercooled water. The base is traditionally at a height between 6,500 and 23,000 ft AGL (26,000 ft in the tropics and 13,000 ft at the poles).

Altocumulus Standing Lenticular (ACSL) Clouds formed at the tips of vertical waves in the wake of mountain ranges.

altostratus A bluish veil or layer of clouds having a fibrous appearance. The outline of the sun may show dimly as if through frosted glass. The base is traditionally at a height between 6,500 and 23,000 ft AGL (26,000 ft in the tropics and 13,000 ft at the poles).

anafront Also called active front. A cold front in which there is a tendency for air to ascend the frontal surface. Generally associated with lift and weather behind the front. (cf. katafront)

anemometer A device that measures wind speed.

ANL Analysis

anomalous propagation Unexpected radio wave propagation that occurs due to non-standard atmospheric conditions. Usually refers to ducting of the beam to the ground, returning ground clutter.

anticyclone An area of high pressure, around which the wind blows clockwise (counterclockwise in the Southern Hemisphere).

anvil The spreading top of a cumulonimbus cloud.

AOA At or above

AOB At or below

AP Anomalous propagation (q.v.)

arctic air Air which has its roots over the snow-covered region of northern Canada, the polar basin, and northern Siberia.

arctic jet Baroclinic jet which develops in association with the polar vortex.

Area Forecast Discussion (AFD) A discussion of the meteorological thinking used in the creation of a zone forecast. (NWS)

ARINC Aeronautical Radio, Incorporated. Company based in Annapolis, MD responsible for

ACARS (q.v.)

ARTCC Air Route Traffic Control Center

As Altostratus (q.v.) Also AS.

ASCII American Standard Code for Information Interchange

ASOS Automated Surface Observing System. The network in place across the United States which have provided automated meteorological reports since 1992.

ATTM At this time

AVA Anticyclonic vorticity advection

AVN 1. The NCEP Aviation model, also known as the global spectral model, comprising the United States' primary global weather model; replaced by the GFS model (q.v.) 2. Aviation.

AWC Aviation Weather Center

AWIPS Advanced Weather Interactive Processing System

AWOS Automated Weather Observing System

back door cold front A cold front moving south or southwestward along and near the Appalachians.

backing Referring to a change in wind direction that is counterclockwise, with respect to either height or time. Contrast with veering.

backscatter Power that returns to the radar dish after striking a target.

baroclinic zone An area in which a horizontal temperature gradient exists. Rapid weather changes may occur in such zones.

barotropic More properly referred to as equivalent barotropic, this term refers to a weather system which has weak or insignificant temperature contrasts

barotropic zone An area in which a significant horizontal temperature gradient does not exist. Rapid weather changes are not as likely as in a baroclinic zone.

Base Reflectivity (BR) A simple reflectivity product as obtained from any elevation of a radar scan (not necessarily the lowest one).

base velocity A simple velocity product as obtained from any elevation of a radar scan (not necessarily the lowest one).

beam width In radar meteorology, the width within which the power density is at least half that of the axis of the beam (i.e. within 3 dB)

blizzard A winter storm which produces, for at least 3 hours, both winds gusting to 35 mph and falling/drifting/blowing snow reducing visibility to less than 1/4 mile.

block A long wave pattern, usually revealed on 200/250/300 mb charts, in which the long waves are neither progressing nor retrogressing. Often refers to the responsible feature, such as an omega or rex block (q.v.)

boundary layer (BL, PBL) The layer in contact with

the ground in which friction is significant. This is usually the lowest few thousand feet of the atmosphere but may vary greatly with weather pattern, season, and insolation.

broken Partial coverage of the sky by a layer of more than half (5 to 7 oktas). (cf. clear, few, scattered, and overcast)

BR Base Reflectivity (q.v.)

BRN Bulk Richardson Number (q.v.)

BUFR Binary Universal Format for Data Representation

Bulk Richardson Number (BRN) The ratio of CAPE to vertical wind shear. It has been found that values of less than 45 support supercellular structures, while greater than 45 favors multicells. However it is not as good of a predictor as its component terms are.

BWER Bounded Weak Echo Region

CAA Cold air advection (q.v.)

cap A layer of warm air aloft that acts as an inversion and suppresses convective development. It may be measured by the Convective Inhibition Index, or CINH (q.v.)

CAPE Convective Available Potential Energy (q.v.)

Cb, CB Cumulonimbus (q.v.)

Cc, CC Cirrocumulus (q.v.)

CCL Convective Condensation Level (q.v.)

Ci, CI Cirrus (q.v.)

CIN Convective inhibition

CISK Convective instability of the second kind

cirrocumulus (Cc) A layer of high, fibrous clouds with convective cells. The cloud is made up entirely of ice crystals. Its bases are traditionally as low as 16,000 ft (20,000 ft in the tropics; 10,000 ft in polar regions).

cirrostratus (Cs) A thin layer of high, fibrous clouds without detail and often appearing as a sheet covering the sky. It is composed entirely of ice crystals. Its bases are traditionally as low as 16,000 ft (20,000 ft in the tropics; 10,000 ft in polar regions).

cirrus (Ci) A layer of high, fibrous clouds composed entirely of ice crystals. Its bases are traditionally as low as 16,000 ft (20,000 ft in the tropics; 10,000 ft in polar regions).

clear Complete absence of cloud. (cf. few, scattered, broken, and overcast)

cloud height The height of a cloud's base, usually rounded to the nearest hundred feet (thousand feet above 10,000 ft).

cold front The leading edge of an air mass that is replacing a warmer air mass.

Composite Reflectivity (CR) A WSR-88D radar product that displays the maximum reflectivity observed in a grid box at a given location.

confluence A pattern in which wind flows inward into a common axis. It is not the same as convergence. (cf. difluence, convergence, divergence)

convection The transport of heat and moisture by the vertical movement of air in an unstable atmosphere. This may cause cumuliform clouds and thunderstorms.

Convective Available Potential Energy (CAPE) the vertically integrated buoyancy of a rising air parcel. Measured in j/kg.

Convective Condensation Level (CCL) The height at which a parcel of air, if heated sufficiently from below, will rise adiabatically until saturation begins.

Convective Inhibition (CIN) A measure of negative buoyancy that prevents a rising parcel from reaching its Level of Free Convection, or LFC (q.v.). It is measured in j/kg.

convective temperature The theoretical surface temperature for a given atmospheric profile that must be reached to start the formation of convective clouds.

convergence A wind pattern in which more air is entering than leaving, either through speed convergence or confluence. (cf. divergence, difluence, confluence)

CONUS Continental United States

Coordinated Universal Time (UTC) See Universal Coordinated Time.

COOSAC Committee on Operations, Standards, and Conventions

Coriolis effect The effect caused by the Earth's rotation which deflects parcels to the right (left in the Southern Hemisphere).

COTR Contract Office Technical Representative

couplet Adjacent maxima of radial velocities of opposite signs.

CPC Climate Prediction Center

CR Composite Reflectivity (q.v.)

cross section A diagram of the atmosphere in which horizontal distance is expressed on the X-axis and height on the Y-axis.

Cross Totals index (CT) An expression of instability, equalling $Td_{850}-T_{500}$. Values of greater than 18-30 are considered significant.

Cs, CS Cirrostratus (q.v.)

CSI Conditional symmetric instability

CT Cross Totals index (q.v.)

Cu, CU Cumulus (q.v.)

cumulonimbus (Cb) A large, cauliflower-shaped cloud whose upper portions are usually fibrous. Often associated with precipitation and thunder.

cumulus (Cu) Low, heaplike clouds that are associated with convective weather. The three "categories" of cumulus are typically fair-weather, moderate, and towering. Further cumulus development will evolve into cumulonimbus.

CVA Cyclonic vorticity advection

cyclogenesis The intensification of a low-pressure system.

cyclone An area of low pressure with a closed circulation. The wind flow rotates counterclockwise (clockwise in the Southern Hemisphere).

dBZ Decibels of reflectivity factor.

decoupling The intensification of the contrast between the boundary layer and the free atmosphere, which strengthens winds above and weakens winds below. Tends to occur at night.

DELMARVA Delaware-Maryland-Virginia

derecho A widespread and fast-moving convective windstorm.

difluence Alternate spelling of diffluence (q.v.)

diffluence A pattern in which wind flows outward from a common axis. It is not the same as divergence. (cf. confluence, convergence, divergence)

diurnal 1. Occurring on a daily basis. 2. Occurring during the day. (cf. nocturnal)

divergence A wind pattern in which more air is leaving than entering, either through speed divergence or diffluence. (cf. convergence, difluence, confluence)

DOCBLOCK Computer program documentation block

dryline A boundary which separates dry, warm continental air from moist, warm oceanic air. It is most common in the Great Plains, the Sahel, India/Bangladesh, Australia, and China.

dynamics A term that generally refers to forces produced by air out of geostrophic balance which in turn produces vertical motion.

easterly wave A disturbance embedded in the trade winds that moves east to west.

EBDIC Extended Binary-Coded Decimal Interchange Code

ECMWF European Centers for Medium Range Weather Forecasting

EHI Energy Helicity Index (q.v.)

EL Equilibrium Level (q.v.)

EMC Environmental Modelling Center (q.v.)

Energy Helicity Index (EHI) An index that is a product of shear and instability, and is defined as CAPE x SRH / 160,000, where CAPE is j/kg and SRH is in m^2/s^2.

Environmental Modelling Center (EMC) A center of NCEP that is focused on improving numerical modelling technologies.

Equilibrium Level (EL) The height, sometimes within the stratosphere, at which a rising parcel's temperature becomes equal to that of the environment. Upward momentum is sharply lost beyond this point.

Eta Eta model

European model The ECMWF global spectral model.

Exit region The region downstream from a jet max. The poleward side typically is associated with divergence aloft and upward motion; the equatorward side with convergence aloft and downward motion.

FA Area forecast

FAA Federal Aviation Administration

Family of Services (FOS) The public connection to National Weather Service data which was established in 1983.

FD Winds and temperatures aloft forecast

FEW Partial coverage by cloud material of a quarter or less (1 to 2 oktas). (cf. clear, scattered, broken, overcast)

FFG Flash flood guidance

FNL Final production run for a given cycle

FNMOC Fleet Numerical Oceanography Center

FNOC Fleet Numerical Oceanographic Center (obsolete; replaced by FNMOC)

FOS Family Of Services (q.v.)

FT Terminal forecast (obsolete; now TAF)

FTP File Transfer Protocol

FTS Federal Telecommunications System

GBL Global production run for a given cycle

GDAS Global Data Assimilation System production run for a given cycle

GDM Graphic Display Model

geostrophic wind The imaginary wind that would result from a balance of both pressure gradient force and the Coriolis effect.

GES Guess

GFS Global Forecast System (q.v.)

Global Forecast System (GFS) The most sophisticated global spectral model currently used by the United States. It incorporates both the AVN and MRF models, whose names have been "retired".

GMT Greenwich Mean Time

GOES Geostationary Operational Environmental Satellite. The United States' network of geostationary weather satellites poised at the Equator above the Western Hemisphere continuously since 1974.

GRIB Gridded Binary data

GTS Global Telecommunications System

HADS Hydrometeorological Automated Data System

HIC Hydrologist In Charge

HMT Hydrometeorological Technician

hodograph A polar coordinate graph showing the wind profile of the atmosphere at a given point, with respect to ground-relative azimuth and speed.

HPC Hydrometeorological Prediction Center (q.v.)

hurricane A warm-core tropical system that has sustained surface winds exceeding 63 kt.

Hydrometeorological Prediction Center (HPC) A center of NCEP which is responsible for cen-

tralized forecasting functions of the National Weather Service.

ICAO International Civil Aviation Organization

IMSL International Mathematical and Statistical Library

instability An atmospheric state in which warm air is able to continue rising and accelerating.

inversion An increase in temperature with height, comprising a stable layer in the atmosphere. Vertical motion through the inversion is suppressed.

INVOF In vicinity of

IR Infrared

isallobar A line of equal atmospheric pressure change.

isentrope A line of equal potential temperature.

isentropic lift Lift produced by motion of air along surfaces of constant potential temperature which slope upward relative to the parcel's motion. This typically occurs when the parcel is traversing from warmer to colder air below.

isentropic subsidence Sinking motion produced by motion of air along surfaces of constant potential temperature which slope downward relative to the parcel's motion. This typically occurs when the parcel is traversing from colder to warmer air below.

isobar A line of equal pressure.

isochrone A line of equal time.

ISPAN Information Stream Project for AWIPS/NOAAPORT

J/KG Joules per kilogram

jet max A region of maximum winds within a jet stream. Also jet streak, speed max.

jet streak A region of maximum winds within a jet stream. Also jet max, speed max.

JIF Job Implementation Form

JMA Japan Meteorological Agency

JSC Johnson Spaceflight Center

JSPRO Joint Systems Program Office for NEXRAD

katafront Also called inactive front. A cold front in which there is a tendency for air to descend the frontal surface. Generally associated with subsidence behind the front and weather ahead of the front. (cf. anafront)

K-Index (KI) A measure of the thunderstorm potential based on vertical temperature lapse rate, moisture content of the lower atmosphere, and the vertical extent of the moist layer. Equals T_{850}-T_{500}+Td_{850}-DD_{700} where DD equals dewpoint depression. Values above 20-35 are significant.

KI K-Index (q.v.)

knot A measure of velocity, nautical miles per hour, equal to 1.15 statute miles per hour.

lapse rate The change in temperature with height. Normally is 6.5 Celsius degrees per km.

LAWRS Limited Aviation Weather Reporting Station (usually a control tower)

LCN Loosely Coupled Network

LI Lifted Index (q.v.)

LCL Lifted Condensation Level (q.v.)

LEWP Line echo wave pattern

LFC Level of free convection

LFM Limited-area Fine Mesh model, a numerical model that was used by NMC (NCEP) from 1971 to 1996.

LFQ Left-front quadrant of a jet streak. In the Northern Hemisphere this is usually associated with upward motion.

LRQ Left-rear quadrant of a jet streak. In the Northern Hemisphere this is usually associated with downward motion.

Lifted Condensation Level (LCL) The height at which a parcel of air will become saturated if lifted adiabatically.

Lifted Index (LI) The temperature difference between a lifted parcel and that of its environment at 500 mb. This is a single-level expression of instability. It equals $T_E - T_P$ where E is the environment and P is the parcel. Negative values are unstable, and below -5 are significant.

LLJ Low Level Jet (q.v.)

Low Level Jet (LLJ) An elongated area of strong winds, generally below 10,000 ft MSL, which may occur in advance of extratropical lows. It is significant in transporting heat and moisture poleward, reinforcing baroclinicity and destabilizing the atmosphere.

long wave A large-scale wave in the upper atmosphere, either a trough or a ridge. There are usually four or five long waves around a hemisphere.

M2/S2 Meters squared per second squared

MAR Modernization and Associated Restructuring Program

MAX Maximum

Maximum Parcel Level (MPL) The highest attainable level a thunderstorm updraft can reach, where all further upward velocity of a parcel is lost. Factors in overshooting tops.

MB Millibars

MCC Mesoscale convective complex

MCD Mesoscale discussion

MCIDAS Man-Computer Interactive Data Access System

MCS Mesoscale Convective System

MDR Manually Digitized Radar (now obsolete)

mesocyclone A low pressure area which is the embodiment of a rotating thunderstorm; it usually measures 1 to 5 miles in diameter. It is a misnomer because it is not a mesoscale system.

mesolow A mesoscale low-pressure area. Not to be confused with mesocyclone.

mesohigh A mesoscale high-pressure area, sometimes associated with stagnating thunderstorm outflow air.

mesoscale Referring to weather systems with scales of about 50 to 500 miles, or 1 to 24 hours.

METAR Meteorological Aviation Report

MIC Meteorologist In Charge

MLCAPE Mean Layer CAPE. Calculated using a parcel that contains mean temperature and mixing ratio of a layer, typically 100 mb deep.

MOA Memorandum of Agreement

Model Output Statistics (MOS) A statistical forecasting model, usually calculated city-by-city.

monsoon A seasonal shift in wind direction.

MOS Model Output Statistics (q.v.)

MOU Memorandum of Understanding

MPL Maximum Parcel Level (q.v.)

MPC Marine Prediction Center

MRF Medium Range Forecast model (obsolete; replaced by GFS)

MSL (above) Mean Sea Level

MSLP Mean Sea Level Pressure

MUCAPE Most Unstable CAPE. CAPE calculated from a parcel that provides the most unstable CAPE possible.

NASA National Aeronautics and Space Administration

National Centers for Environmental Prediction (NCEP) Was NMC (National Meteorological Center) from 1958-1995. An agency falling under NOAA which provides guidance and products to the National Weather Service. It is comprised of nine centers: Aviation Weather Center (AWC); Climate Prediction Center (CPC); Environmental Modelling Center (EMC); Hydrometeorological Prediction Center (HPC); NCEP Central Operations (NCO); Ocean Prediction Center (OPC); Space Environmental Center (SEC); Storm Prediction Center (SPC); and Tropical Prediction Center (TPC).

National Climatic Data Center (NCDC) The United States government agency responsible for archival of meteorological data.

National Oceanic and Atmospheric Administration (NOAA) The United States government agency falling under the Department of Commerce, which is responsible for all civilian programs engaged in work with the atmosphere, oceans, and lakes.

National Weather Service (NWS) A branch of NOAA responsible for all United States public forecasting.

NCCF NOAA Central Computer Facility

NCDC National Climatic Data Center (q.v.)

NCEP National Centers for Environmental Prediction (q.v.)

NCO NCEP Central Operations

negative tilt Description of an upper-level trough whose axis is tilted to the west with increasing latitude. It is often associated with strengthening dynamics.

NESDIS National Environmental Satellite Data and Information Service

NEXRAD Next Generation Weather Radar (WSR-88D)

NEXUS Next Generation Upper-Air System

NGM Nested Grid Model

NHC National Hurricane Center

NIDS NEXRAD Information Dissemination Service

nimbostratus An amorphous cloud thick enough to completely obscure the sun, with its base almost indistinguishable and typically obscured by precipitation. Does not produce showers or thunder. Abbreviated Ns.

NMC National Meteorological Center (obsolete; now NCEP)

NMFS National Marine Fisheries Service

NOAA National Oceanic and Atmospheric Administration (q.v.)

NOS National Ocean Survey

NOTAM Notice to Airmen

nowcast A forecast of about six hours or less. Also called a short-term forecast.

Ns Nimbostratus (q.v.)

NS Nimbostratus (q.v.)

NSSFC National Severe Storms Forecast Center

NVA Negative vorticity advection. Advection of negative vorticity into a region.

NWS National Weather Service (q.v.)

NWSTG National Weather Service Telecommunications Gateway

occlusion The convergence of three air masses, in which the least dense is displaced aloft and the remaining two are demarcated by an occluded front. Typically occurs when a cold front "catches up" to a warm front.

OFOAR Office of Oceanic and Atmospheric Research

OHP One-Hour Precipitation, as used in weather radar estimates.

OI Optimum Interpolation method

okta An eighth of sky cover.

omega block An upper-level pattern in which a high pressure (height) area intensifies to a very high amplitude, resembling the greek letter omega. It "locks in" the long wave pattern.

ON Office Note

outflow boundary The leading edge of outflow from a thunderstorm downdraft. It may persist hours or days after the dissipation of the storm.

overcast A cloud layer completely covering the sky (8 oktas of cover). (cf. clear, few, scattered, and broken)

overrunning An oversimplification of the process of isentropic lift (q.v.).

PE Primitive Equation model

PFJ Polar front jet (q.v.)

PIREP Pilot Report

polar front jet (PFJ) The jet that is associated with the gradient between polar and tropical air masses. (cf. arctic jet, subtropical jet, low-level jet)

polar vortex A large cold-core low aloft that typically is found over northern Hudson Bay in North America during the winter months. Occluding polar front systems are usually absorbed into the polar vortex.

POP Probability of Precipitation (q.v.)

positive area The area formed on a sounding between an environmental temperature line and a warmer parcel temperature line. Its area is roughly proportional to CAPE.

positive-tilt Description of an upper-level trough whose axis is tilted to the east with increasing latitude. It is often associated with weakening dynamics.

potential temperature The temperature which a parcel would have if brought to a common level, by standard convention 1000 mb.

pressure gradient The change in pressure over a given distance.

Probability of Precipitation (POP) A quantity that describes the likelihood of a measurable amount of precipitation at any given location in a forecast area. The NWS expressions are 20% for slight chance, 30-50% for a chance, and 60-70% for likely.

PROD Production (for operational jobs)

profiler Also wind profiler. A radio detection device designed to measure wind direction and speed vertically in the troposphere above a given point.

PVA Positive vorticity advection. Equal to CVA in the Northerm Hemisphere and AVA in the Southern Hemisphere (q.v.)

Q vector A horizontal vector representing the rate of change of the horizontal potential temperature gradient. Convergence or divergence of the vectors are associated with forcing for vertical motion.

QG Quasi-geostrophic

QPF Quantitative Precipitation Forecast

RAOB Radiosonde observation

radial velocity The component of motion along an axis extending from a radar unit. The NEXRAD base velocity product depicts radial velocity.

RAFS Regional Analysis and Forecast System (NGM)

range folding A process by which a radar echo returns after another pulse has been transmitted, creating an echo that might be incorrectly distanced by the radar unit.

RAREP Radar Report

RCM Radar Coded Message. An automated product of the WSR-88D unit which provides a summary of the echoes and signatures from a given radar.

rex block A blocking pattern in the upper levels in which a closed high is located poleward of a closed low. The long-wave pattern tends to "lock up".

RFQ Right-front quadrant of a jet streak. In the Northern Hemisphere this is usually associated with downward motion.

RGL Regional Model

ridge An elongated area of high pressure or heights.

RRQ Right-rear quadrant of a jet streak. In the Northern Hemisphere this is usually associated with upward motion.

RUC Rapid Update Cycle model

RUNHIST Run History

SBCAPE Surface based CAPE; resulting from a parcel that is lifted from the surface with no other modifications.

Sc Stratocumulus (q.v.)

SC Stratocumulus (q.v.)

scattered Partial coverage of a cloud layer, covering more than a quarter to half of the sky, of 2 to 4 oktas. (cf. clear, few, broken, and overcast)

SD 1. Radar Report (now obsolete) 2. Storm Data, a publication of NCDC.

SDM Senior Duty Meteorologist

SEC Space Environment Center

Showalter Stability Index (SSI) The difference in temperature between the environment at 500 mb and a parcel lifted from 850 mb, expressed as T_{500}-T_{850}. A negative value corresponds to instability. Lifted Index and CAPE are usually preferred over the SSI.

SIGMET Significant Weather bulletin for pilots

SOO Science and Operations Officer

sounding A plot of temperature and dewpoint above a given station with respect to temperature (X-axis) and height (Y-axis). A thermodynamic diagram, usually the SKEW-T log P, is typically used.

SPC Storm Prediction Center (q.v.)

spectrum width The variance in velocity of scatterers within a given volume of air.

speed max A region of maximum winds within a jet stream. Also jet max, jet streak.

SPENES NESDIS satellite precipitation estimate

SRH Storm-relative helicity

SSI Showalter Stability Index (q.v.)

St Stratus (q.v.)

ST Stratus (q.v.)

STJ Subtropical jet (q.v.)

STK Storage Technology

Storm Prediction Center (SPC) A branch of NCEP, located in Norman, Oklahoma, which is responsible for providing short-term forecast guidance for convective storms.

stratocumulus A relatively flat, low cloud with little vertical development. It has distinct globular masses or rolls.

stratus (St) A low, sheetlike cloud which may either occur alone, or with precipitation (in which case it is referred to as scud, fractus, or stratus of bad weather).

subsidence Sinking motion.

sub-synoptic Mesoscale.

subtropical jet (STJ) An upper-level jet stream that is usually found between 20 and 30 deg of latitude and is associated with thermal differences within the subtropical high. (cf. arctic jet, polar front jet, and low-level jet)

SWODY1 Severe Weather Outlook - Day 1

SWODY2 Severe Weather Outlook - Day 2

synoptic-scale Spanning a distance scale of over 500 miles or a time scale of days.

TAF Terminal Aerodrome Forecast

TCU Towering Cumulus

TD Tropical Depression

teleconnection A strong statistical relationship between weather in different parts of the globe.

theta-e Equivalent potential temperature

Total Totals Index (TTI) A sum of the Cross Totals and Vertical Totals indices. It is equal to T_{850}-T_{500}+Td_{850}-T_{500}. A value of greater than 44-56 is considered significant.

TPB Technical Procedures Bulletin

TPC Tropical Prediction Center (q.v.)

Tropical Prediction Center (TPC) A branch of NCEP responsible for tropical weather forecasting, including hurricanes.

tropical storm A warm-core storm with a maximum sustained surface wind of 34-63 kt.

tropopause The point between the troposphere and stratosphere at which a positive tropospheric lapse rate becomes neutral or negative.

trough An elongated area of low pressure or heights.

TS Tropical Storm

TTI Total Totals Index (q.v.)

typhoon A tropical storm of hurricane strength in the Western Pacific basin.

UA Pilot Report

UCAR University Corporation for Atmospheric Research

UCL UNICOS Control Language (shell script)

UKMO United Kingdom Met Office

UKMET United Kingdom Met Office

ULJ Upper level jet

UPS Uninterruptable Power Supply

UTC Universal Coordinated Time

UVV Upward vertical velocity

VAD Velocity Azimuth Display. A plot of radial velocity (Y-axis) with respect to azimuth (X-axis) by a weather radar for a given level. It is used as a basis for construction of VWP diagrams (q.v.)

VAFTAD Volcanic Ash Forecast Transport and Dispersion

VC Vicinity

veering Referring to a clockwise change in the wind direction, with respect to either height or time. Contrast with backing.

vertical stack The tendency for a weather system, usually a closed low or high, to have the same location aloft as at the surface. This typically indicates a lack of baroclinicity and suggests a warm-core or cold-core structure.

Vertical Totals index (VT) An expression of the low to mid-level lapse rate, as given by T_{850}-T_{500}. A value of 26 or more is considered significant.

VIL Vertically Integrated Liquid

volume scan The complete scan of a weather radar for all assigned elevations. When a volume scan is complete, the radar is able to generate all possible products (with the exception of products that require a history of an echo). The WSR-88D completes a volume scan in 5 to 10 minutes.

vort max The highest vorticity in a given region.

vorticity The rotation in a volume of air, made up of shear and curvature.

VSB Visible

VIS Visible

VWP VAD Wind Profile. A plot of winds with height above a given station, as determined by a weather radar. The profile is displayed with height as the Y-coordinate and time as the X-coordinate.

VT Vertical Totals index (q.v.)

WAA Warm air advection

WAFS World Area Forecast System

WBZ Wet Bulb Zero (q.v.)

Wet Bulb Zero (WBZ) The height at which the wet bulb temperature drops below freezing, expressed as height above ground level (AGL). It is a measure of depth through which a hailstone will melt. Values of less than 10,000 ft are associated with large hail, given enough instability.

WFO Weather Forecast Office

WMSC Weather Message Switching Center

WMO World Meteorological Organization

WS Significant Weather bulletin (SIGMET) for pilots

WSFO Weather Service Forecast Office

WSO Weather Service Office

WST Convective SIGMET for pilots

WW Weather watch (thunderstorm or tornado)

WWB World Weather Building

Z Zulu Time (Greenwich Mean Time)

ZFP Zone Forecast Product (q.v.)

Zone Forecast Product (ZFP) A NWS bulletin that provides a clear, chronological statement of the weather conditions in a county or a given set of counties for the general public.

Suggested Internet Weather Sites

Provided below are some excellent launching points for finding even more weather fore-casting charts and diagrams. Have suggestions for the next edition? We'd love to hear about them. E-mail the author at the address listed in the introduction.

❑ Original Sources

When you want an original source of data from a site that actually hosts the graphics, here's where to start.

<www.nco.ncep.noaa.gov/pmb/nwprod/analysis>
NCEP Central Operations / Product Management Branch — Excellent collection of the ETA, GFS, andNGM runs, direct from one of the biggest modelling centers in the world.

<weather.cod.edu>
College of DuPage — The College of DuPage, just west of Chicago, has some of the sharpest minds in operational forecasting as well as a cutting-edge data site to boot.

<www.rap.ucar.edu/weather>
UCAR Real-Time Weather Data — Managed by Greg Thompson at the National Center for Atmo-spheric Research, this site has been a longtime favorite of many weather hobbyists.

<twister.sbs.ohio-state.edu>
Ohio State University — The OSU weather site is dependable, thorough, and packed with graphics.

❑ Links and Data Listings

Sometimes it can be better to set your browser's start page to a wider variety of data rather than just concentrating on one source. Here are some good options to look through.

<www.stormeyes.org/tornado/rogersif.htm>
Roger Edwards' Storm Intercept Forecasting Links — Not just for storm forecasting!

<www.weathercharts.org>
David Hayfield's Europe weather site — Lots of links covering Europe.

<www.westwind.ch>
Westwind — A very extensive list of links for European weather

Suggested References and Further Reading

This book is not intended to be authoritative on any of the material presented. Therefore presented here is a list of source materials for this book as well as useful, comprehensive resources that can be sought after for further reading. Internet URL's are provided where applicable.

General forecasting education and reference

Ahrens, C. Donald (1994). Meteorology Today: An Introduction to Weather, Climate, and the Environment. West Publishing Co., St. Paul (ISBN 0-314-02779-3). 592 pp.

Carlson, Toby N. (1991). Mid-Latitude Weather Systems. Routledge, London (ISBN 0-415-10930-2). 507 pp.

Cole, Franklyn W. (1980). Introduction to Meteorology. John Wiley & Sons, New York (ISBN 0-471-04705-8). 505 pp.

Gedzelman, Stanley D. (1980). The Science and Wonders of the Atmosphere. John Wiley & Sons, New York (ISBN 0-471-02972-6). 535 pp.

Moran, Joseph M. (1994). Meteorology: The Atmosphere and the Science of Weather. Macmillan, Englewood Cliffs (ISBN 0-02-383341-6). 517 pp.

National Weather Service (1993). Forecasters Handbook No. 1. 340 pp.

Stull, Roland B. (1995). Meteorology Today for Scientists and Engineers. West Publishing Co., St. Paul (ISBN 0-314-06471-0). 385 pp.

Vasquez, Tim (2001). Weather Forecasting Handbook. Weather Graphics Technologies, Garland (ISBN 0-9706840-2-9). 204 pp.

Observational charts

Doswell, Charles A. III (1986). The human element in weather forecasting. Nat. Wea. Dig., 11, 6-18. <www.cimms.ou.edu/~doswell/human/Human.html>

Djuric, Dusan (1994). Weather Analysis. Prentice-Hall, Englewood Cliffs (ISBN 0-13-501149-3). 304 pp.

Miller, Robert C. (1972). Notes on Analysis and Severe-Storm Forecasting Procedures of the Air Force Global Weather Central. AWS Technical Report 2000 (Rev), Air Weather Service, Scott AFB. 190 pp.

National Weather Service (1993). Graphical Guidance. National Weather Service, Washington. 169 pp.

Sanders, Frederick and Doswell, Charles A. III (1992). A Case for Detailed Surface Analysis. Bull. of the Amer. Met. Soc., 76, 505-521.

Young, G. S. and Fritsch, J. M. (1989). A Proposal for General Conventions in Analyses of Mesoscale Boundaries. Bull. of the Amer. Met. Soc., 70: 1412-1513. <ams.allenpress.com>

Satellite Imagery

Cooperative Program for Operational Meteorology, Education, and Training (COMET). COMET Satellite Meteorology Course: Meteorological Sounders. University of Wisconsin homepage. <cimss.ssec.wisc.edu/goes/comet/sounder.html>

National Environmental Satellite, Data, and Information Service (1983). The GOES Users Guide, NESDIS, Washington. 164 pp.

NOAA Satellite and Information Services. NOAASIS homepage. <noaasis.noaa.gov/NOAASIS>

Schmit, Timothy J., Wade, Gary S., and Aune, Robert M. (1998). Automated GOES Sounder Products. Proceedings of the Satellite Applications Conference, March 4-6, 1993, Asheville. <cimss.ssec.wisc.edu/goes/sounder/products.html>

Numerical Models

Allen, C., Kramer, D., Smith, R., and Stults, A. (2001). Vertical Resolution and Coordinates. Texas A&M University homepage. <www.met.tamu.edu/class/metr452/models/2001/vertres.html>

Canadian Meteorological Center (2003). GEM: The Global Environmental Multiscale Model. CMC homepage. <www.cmc.ec.gc.ca/rpn/gef_html_public>

Carr, Frederick H. (1988). Introduction to Numerican Weather Prediction Models at the National Meteorological Center. 63 pp.

Evenson, Eric C. and Strobin, Mark H. (1998). Model Boundary Layer Problems and Their Impact on Thunderstorm Forecasting in the Western United States. NWS WR Technical Attachment TA 98-20. <www.wrh.noaa.gov/wrhq/98TAs/9820>

Fleet Numerical Meteorology and Oceanography Center (2002). Model Characteristics and Tendencies for NOGAPS 4.0, COAMPS 3.0, and WW3 1.18. FNMOC homepage. <https://www.fnmoc.navy.mil/PUBLIC/MODEL_REPORTS/MODEL_TENDENCY_REVIEW/tendencies.html>

Hydrometeorological Prediction Center (2003). Model Biases. NCEP homepage. <www.hpc.ncep.noaa.gov/mdlbias/biastext.shtml>

Japan Meteorological Agency (2002). Outline of the Operational Numerical Weather Prediction at the Japan Meteorological Agency. Appendix to WMO Numerical Weather Prediction Progress Report. <www.jma.go.jp/JMA_HP/jma/jma-eng/jma-center/nwp/outline-nwp/index.htm>

National Meteorological Center (1987). Section 2.2.1: The NMC Production Suite. NMC Handbook., National Meteorological Center, Washington.

National Weather Service, Binghamton Office (2003). Information on Operational Models. NWS Binghamton homepage. <www.erh.noaa.gov/er/bgm/models.htm>

Staudenmaier, Mike, Jr. (1997). The Navy Operational Global Atmospheric Prediction System (NOGAPS). NWS WR Technical Attachment TA97-09. <www.wrh.noaa.gov/wrhq/97TAs/TA9709/ta97-09.html>

Toth, Zoltan and Kalnay, Eugenia (1997). Ensemble Forecasting at NCEP and the Breeding Method. Mon. Wea. Rev.: 125, 3297-3319. <www.atmos.umd.edu/~ekalnay/TothKalnay97.pdf>

United Kingdom Met Office (2003). NWP Gazette, quarterly. <www.metoffice.gov.uk/research/nwp/publications/nwp_gazette>

University Corporation for Atmospheric Research (2003). Operational Models Matrix: Characteristics of Operational NWP Products. UCAR Homepage. <meted.ucar.edu/nwp/pcu2>

Weickman, Klaus, Whitaker, Jeff, Roubicek, Andres, and Smith, Catherine. The use of ensemble forecasts to produce improved medium range (3-15 days) weather forecasts. NCEP Homepage. <www.cdc.noaa.gov/spotlight/12012001>

Text

Office of the Federal Coordinator for Meteorology (1998). Federal Meteorological Handbook #1: Surface Weather Observations and Reports (FCM-H1-1998). OFCM, Washington. <www.ofcm.gov/fmh-1/fmh1.htm>

Office of the Federal Coordinator for Meteorology (1997). Federal Meteorological Handbook #3: Rawinsonde and Pibal Observations (FCM-H3-1997). OFCM, Washington. <www.ofcm.gov/fmh3/text/default.htm>

Office of the Federal Coordinator for Meteorology (1998). Federal Meteorological Handbook #12: United States Meteorological Codes and Coding Practices (FCM-H12-1998). OFCM, Washington. <www.ofcm.gov/fmh12/frontpage.htm>

Radar

Allen, S. (1996). Impacts of Optimum Slant Range on WSR-88D VAD Wind Profiles. NWS Houston homepage. <www.srh.noaa.gov/ftproot/HGX/HTML/projects/vad2/main.htm>

Office of the Federal Coordinator for Meteorology (2003). Part A: System Concepts, Responsibilities, and Procedures. Federal Meteorological Handbook #11: WSR-88D Doppler Radar Meteorological Observations (FCM-H11A-2003), OFCM, Washington. <www.ofcm.gov/fmh11/fmh11.htm>

Collins, W. G. (2000). The Quality Control of Velocity Azimuth Display (VAD) Winds at the National Centers for Environmental Prediction. Preprints, 11th Symposium on Meteorological Observations and Instrumentation, Albuquerque, NM, Amer. Met. Soc., 317-320. <www.emc.ncep.noaa.gov/mmb/papers/collins/preprints/vadqc.htm>

U. S. Air Force (1992). WSR-88D Products. Study Guide C40ST2524 009-SW-101C, Chanute Training Center, Chanute AFB. 42 pp.

Webber, Richard D. (1996). Forecasting Turbulence and Icing using the WSR-88D VAD Wind Profile Product. NWS CR Applied Research Paper ARP20-11. <www.crh.noaa.gov/techpapers/arp20/20-11.html>

Miscellaneous

Beran, D.W. and Wilfong, T.L. (1998). U.S. Wind Profilers: A Review (FCM-R14-1998). Office for the Federal Coordinator of Meteorology, 56 pp. <www.ofcm.gov/r14/front.htm>

United Kingdom Met Office (2003). CWINDE Network. Met Office homepage. <www.metoffice.gov.uk/research/interproj/cwinde>

World Meteorological Organization (1988). Manual on Codes. WMO Publication No. 306, World Meteorological Organization, Geneva

Index